Practical Biochemistry

Edited by

Pamela Jha
Department of Biological Sciences
Sunandan Divatia School of Science SVKM'S NMIMS
Deemed to be University
Vile-Parle (West), Mumbai, India

Practical Biochemistry

Editor: Pamela Jha

ISBN (Online): 978-981-5165-85-2

ISBN (Print): 978-981-5165-86-9

ISBN (Paperback): 978-981-5165-87-6

First published in 2024.

need for a court order if at any point you breach any terms of this License Agreement. In no event will any delay or failure by Bentham Science Publishers in enforcing your compliance with this License Agreement constitute a waiver of any of its rights.

3. You acknowledge that you have read this License Agreement, and agree to be bound by its terms and conditions. To the extent that any other terms and conditions presented on any website of Bentham Science Publishers conflict with, or are inconsistent with, the terms and conditions set out in this License Agreement, you acknowledge that the terms and conditions set out in this License Agreement shall prevail.

Bentham Science Publishers Pte. Ltd.
80 Robinson Road #02-00
Singapore 068898
Singapore
Email: subscriptions@benthamscience.net

BENTHAM
SCIENCE

CONTENTS

FOREWORD

Biochemical methods have contributed to diverse fields of science and technology over a century. They continue to be even more relevant today, half a century after molecular biology, genomics, and bioinformatics seem to hog the limelight. After all the genes are cloned and all the genomes are sequenced and analyzed bioinformatically, there is still no substitute for wet biochemistry to make progress in any area of modern biology or biotechnology. Apart from helping us in understanding the biological processes in any living organism, biochemical methods are critical in many applied branches such as medicine, agriculture, industrial microbiology, and biotechnology. Therefore, rigorous training in the protocols and practical skills of biochemistry greatly enhances the academic success and employability of students in every area of life sciences and biotechnology. This necessitates constant upgradation of books and manuals, and this book is the latest effort in that direction.

"Practical Biochemistry" edited by Dr. Pamela Jha is a comprehensive guide designed for students and early researchers in the whole gamut of life sciences. It encompasses various disciplines such as cell biology, molecular biology, and medical sciences, offering a holistic understanding of biochemistry through its underlying methods. The book rightly emphasizes both theoretical foundations and practical protocols, catering to individuals at the bachelor's and master's levels, as well as research scholars. Each experiment is accompanied by a quick recap of the relevant theory, including specific reactions and protocols, supplemented by 'viva voce questions' for interactive learning. Expected results with quantitative values are provided for many experiments, aiding in self-verification and comprehension. Dr. Pamela Jha's book is structured for quick and effective revision, making it a valuable resource for students and enthusiasts alike. Its user-friendly approach and comprehensive coverage ensure an enjoyable learning experience in the field of biochemistry.

I congratulate Dr. Jha and her team of authors and the publisher for this book.

N. Raghuram
University School of Biotechnology
Guru Gobind Singh Indraprastha University
Dwarka, New Delhi- 110078, India

PREFACE

The first edition provides information that is important for understanding of some basic and advanced biochemical methods for qualitative and quantitative estimation of major macromolecules. This book is organized into 5 sections, *i.e.*, carbohydrates, proteins, nucleic acids, lipids, and other techniques and SDS. Each section comprises multiple chapters structured with abstract, keywords, introduction, aim, principle, requirements, procedure, probable dummy results, conclusion, precautions, and finally, references. In addition, each chapter is also supplemented with relevant interesting facts in the 'did you know' segment and 'viva voice' questions. I have attempted to write this book in a simple and lucid language to enable easy, focused learning.

I welcome constructive comments from students and readers. Finally, I express my gratitude to the contributors and publisher who have extended tremendous support in the production of this first edition. I wish everyone happy learning!

Pamela Jha
Department of Biological Sciences
Sunandan Divatia School of Science SVKM'S NMIMS
Deemed to be University, Vile-Parle (West), Mumbai, India

List of Contributors

Anshika Sah	Environmental Studies and Resource Management, TERI School of Advanced Studies, New Delhi - 110 070, India
Ankita Nitin Padhye	Food Engineering and Technology Department, Institute of Chemical Technology, Mumbai, Maharashtra, India
Ahiree Ghosal	Zydus School for Excellence, Ahmedabad, Gujarat, India
Mrittika Sarkar	Department of Biomedical Engineering, University of Michigan, Michigan, USA
Niriksha Patel	Symbiosis International University, Pune, Maharashtra, India
Payal Agarwal	Graduate School of Medical Science and Engineering, Korean Advanced Institute of Science and Technology, Daejoeon, South Korea
Pranali Bhosale	Environmental Engineering, Institute of Environmental Engineering, National Sun Yat-Sen University, Kaohsiung, Taiwan
Pranali Garde	Stability Data Analytics, Apotex Research Pvt. Ltd., Navi Mumbai, Maharashtra, India
Revati Supekar	Biotechnology, Department of Management Studies, DY Patil University, Pune, India
Sai Joshi	CSIR-NCL, Dr. Homi Bhabha Road, Pune - 411 008, India
Sinchan Hait	Department of Chemical Engineering, BITS-Pilani K.K. Birla Goa Campus, Goa-403726, India

CHAPTER 1

Introduction to Carbohydrates

Mrittika Sarkar[1], **Sinchan Hait**[2,*] and **Sai Joshi**[3]

[1] *Biomedical Engineering, University of Michigan, Michigan, USA*

[2] *Department of Chemical Engineering, BITS-Pilani K.K. Birla Goa Campus, Goa-403726, India*

[3] *CSIR-NCL, Dr. Homi Bhabha Road, Pune - 411 008, India*

Abstract: Carbohydrates are one of the four major classes of biologically essential organic molecules in living organisms. They are the most abundant class of biomolecules in nature based on their mass. And they make up most of the organic bodies on earth due to their extensive role in every life form. Carbohydrates are polyhydroxy aldehydes or ketones or substances that result in such compounds upon hydrolysis. These macromolecules are composed of carbon (C), hydrogen (H), and oxygen (O). They are also referred to as saccharides *(sakcharon = sugar or sweetness)* since most of them have a sweet taste. Apart from that, carbohydrate serves as a primary energy source, a structural framework for nucleotides, and provides structural support to organisms. They also play a crucial role as mediators in cellular interaction. Carbohydrates are classified into monosaccharides, oligosaccharides, and polysaccharides. Monosaccharides are the simplest form of carbohydrates, and a few monosaccharides hydrolyse to form oligosaccharide, and many monosaccharides together form polysaccharides. They also act as a precursor for glycoproteins and glycolipids. In this section, we are discussing some basic concepts about carbohydrates.

Keywords: Carbohydrates, Cellular interaction, Energy, Glycoproteins, Glycolipids, Monosaccharide, Nucleotides, Oligosaccharide, Polysaccharides, Saccharides.

INTRODUCTION

Carbohydrates are the most abundant biomolecule on earth as they make up most of the organic matter. They are widely present in both plant and animal tissues, from serving as skeletal structures in plants, insects, and crustaceans to food reserves in the storage organs of plants and animals. A few carbohydrates (sugar and starch) are major chief dietary components in most parts of the world [1]. The

* **Corresponding author Sinchan Hait:** Department of Chemical Engineering, BITS-Pilani K.K. Birla Goa Campus, Goa-403726, India; E-mail: sinchan.hait1@gmail.com

metabolism of carbohydrates serves as a primary energy source for non-photosynthetic cells. They also serve as energy storage and metabolic intermediate. Deoxyribose and ribose sugar form a structural framework for DNA and RNA [2]. In plant cells and bacterial cells, polysaccharides are an essential element for cell wall formation. Cellulose, which is a key component of the plant cell wall, is the most abundant carbohydrate. In addition, carbohydrates are also linked with proteins and lipids, where they act as precursor molecules for the formation of glycoproteins and glycolipids.

Classification of Carbohydrates

Carbohydrates are mainly classified into 3 groups:

Monosaccharides (mono = one)

They are often called simple sugars and cannot be hydrolyzed into smaller units. They possess a free aldehyde (—CHO) or ketone (=CO) group and 2 or more hydroxyl (—OH) groups. They are represented by the general formula $C_n(H_2O)_n$ or $C_nH_{2n}O_n$ and can be subdivided into trioses, tetroses, pentoses, hexoses, heptoses *etc*., depending upon the number of carbon atoms they possess [3]. They are also classified as aldoses or ketoses, depending upon whether they contain aldehyde or ketone groups. D-glucose is the most abundant monosaccharide found in nature and is a 6- carbon sugar.

Oligosaccharides (oligo = few)

These are sugar compounds that can be hydrolysed to yield 2 to 10 molecules of the same or different monosaccharides. A disaccharide (general formula = $C_n(H_2O)_n - 1$) is an oligosaccharide that yields 2 molecules of monosaccharide on hydrolysis and one yielding 3 molecules of monosaccharide as a trisaccharide (general formula=$C_n(H_2O)_n - 2$) and so on. Examples of oligosaccharides are shown in Table **1.1**.

Table 1.1. Classification of oligosaccharides.

Oligosaccharides	No. of Sugar Molecules	Examples
Disaccharides	2	Sucrose, Lactose
Trisaccharides	3	Rhamnose, Gentianose
Tetrasaccharides	4	Stachyose, Scorodose
Pentasaccharides	5	Verbascose

Polysaccharides (poly = many)

These are sugar compounds that yield more than 10 molecules of monosaccharides on hydrolysis and have a general formula of $(C_6H_{10}O_5)_x$. They are further classified based on monosaccharide molecules produced after hydrolysis.

Homopolysaccharides

Polysaccharides yielding or made up of the same type of monosaccharides. *E.g.*, Starch, Glycogen.

Heteropolysaccharides

Polysaccharides yielding or made up of different monosaccharides. *E.g.*, Hyaluronic acid, Chondroitin.

GLYCOSIDIC BOND

A glycosidic bond is a covalent bond that joins the hemiacetal group of a saccharide or sugar molecule to one another and the hydroxyl group of another compound, like alcohol. This bond is not always a carbohydrate and is mainly formed as a result of a condensation reaction between alcohol or the hydroxyl (-OH) group of one group to the anomeric carbon of the sugar in which water molecule is released out and O-linkage is formed between the two molecules. Anomeric carbon is defined as the carbon atom to which the carbonyl (aldehyde or ketone) group is attached [4].

There are also types of glycosidic bonds and they are classified as:

N-glycosides

In which linkage takes place *via* a nitrogen atom, also known as glycosylamines.

C-glycosides

In which linkage takes place *via* a carbon; this type of linkage is said to be resistant to hydrolysis.

S-glycosides

In which linkage takes place *via* a sulfur atom, also known as thioglycosides.

O-glycosides

In which the linkage takes place *via* an oxygen atom. This type of linkage takes place between sugar molecules and this linkage is also known as ether bonds as ether is an oxygen atom bonded to two carbon atoms which are unreactive and stable compared to other chemical groups, such as alcohol.

REDUCING AND NON-REDUCING SUGARS

Carbohydrates or sugars that have the carbonyl group have free aldehyde (at C-1) or ketone (at C-2), are called reducing sugars. These sugars exist mainly in hemiacetal or hemiketal form and exhibit mutarotation (change in optical rotation), as shown in Fig. (**1.1-A**). Monosaccharides (*e.g.*, Glucose *etc.*) and a few disaccharides (*e.g.*, Lactose, maltose, *etc.*) are examples of reducing sugars (Fig. **1.1-B**) [1, 2].

Non-reducing sugars are carbohydrates that do not have free aldehyde or ketone groups, but they are present in bond formation (Fig. **1.1-C**). They are kind of "closed" sugars and exist mostly in acetal or ketal forms and they do not exhibit mutarotation like reducing sugars [1]. Mainly all polysaccharides and some disaccharides (*e.g.*, Sucrose *etc.*) are examples of these sugars.

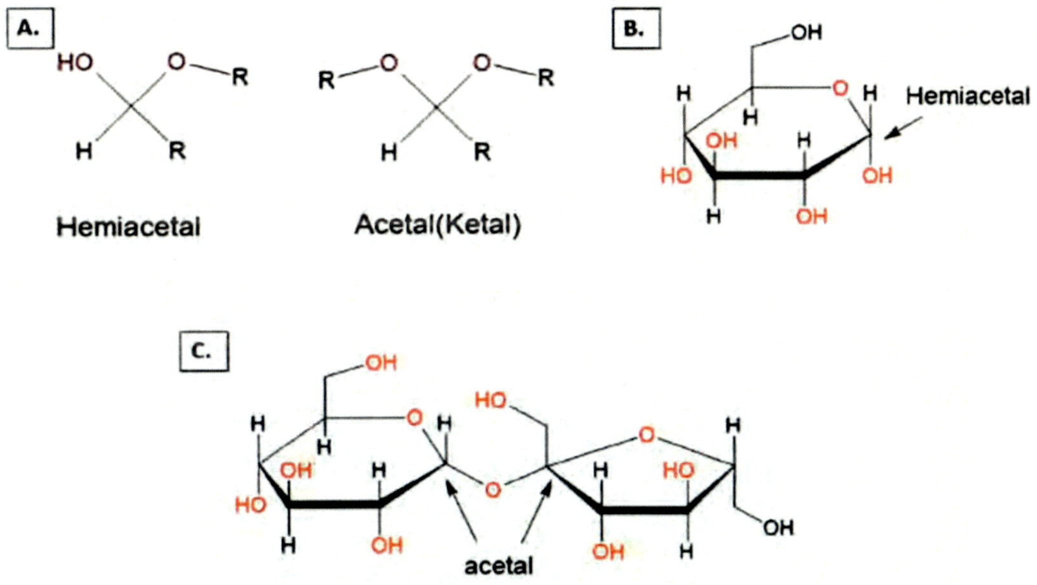

Fig. (1.1). (**A**) Hemiacetal and Ketal structure. (**B**) α-D glucose (Reducing sugar). (**C**) Sucrose (Non-reducing sugar).

PROPERTIES OF CARBOHYDRATES

Elemental Composition

Mainly comprises carbon, hydrogen and oxygen (as the name suggests - carb + hydrates) but can also contain phosphorus (*e.g.*: glycogen), sulphur (*e.g.*: thioglycosides), and nitrogen (*e.g.*: N-acetyl glucosamine). General formula: $C_x(H_2O)_y$; however, there are a few carbohydrates that are an exception to this property (*e.g.*: rhamnose -$C_6H_{12}O_5$) [1 - 3].

Physical Properties

Stereoisomers

Molecules that have the same molecular formula and structure but differ only in the three-dimensional orientation of their constituent atoms in space. They are of 2 types: Geometrical and Enantiomers [1, 2].

Geometrical (cis-trans isomers)

Stereoisomers that have double bonds within their carbon chain whose functional groups are in different spatial configurations, *i.e.*, are either on the same side (*cis*) or across each other (*trans*). *E.g*,, Maleic and fumaric acids (Fig. **1.2**).

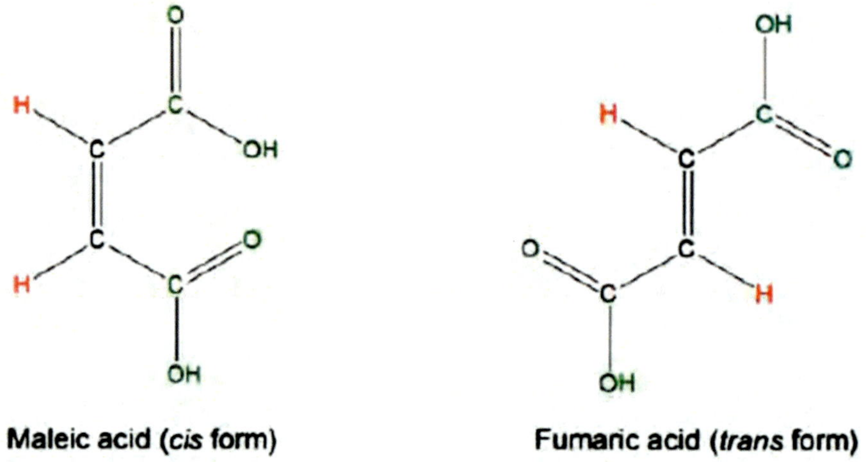

Fig. (1.2). Geometric stereoisomers.

Enantiomers

Stereoisomers, which are non-superimposable mirror images of each other. In carbohydrates, their stereochemistry is designated as D- and L- enantiomers. When the -OH group is on the right-hand side of the penultimate asymmetric carbon in the open chain structure, it is denoted as D-sugar (*E.g.*: D-Arabinose) and if on the left, it is called L-sugar (*E.g.*: L-Arabinose) (Fig. **1.3**).

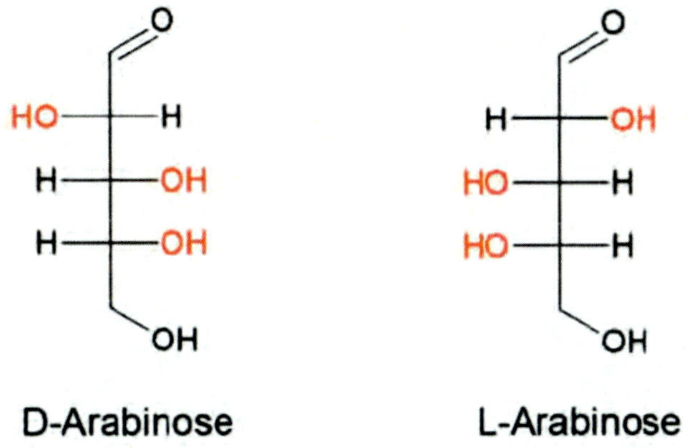

Fig. (1.3). Enantiomers of the sugar arabinose.

Optical Activity

Carbohydrates are denoted as + (or *d*) or - (or *l*) based on their ability to rotate plane-polarized light to right or left, respectively. For *e.g.*: D-Glucose is dextrorotatory and D- Fructose levorotatory and are written as D (+) Glucose or D (d) Glucose D (-) Fructose or D (l) Fructose.

Epimers

These are stereoisomers that differ at one carbon atom only. They have at least two chiral centres but have opposite configurations at only one of the stereogenic centres.

Anomers

These are epimers that differ from each other in the configuration of C1 (aldose) or C2 (ketose). These C1 or C2 carbon atoms are called anomeric or epimeric carbon. The difference lies in whether the hydroxyl group of the anomeric carbon, is present below the plane of the ring structure, in which case it is denoted with α- (Eg: α-D-Fructofuranose), and if present above the plane the sugar is said to be β-form (Eg: β-D-fructofuranose) see Fig. (**1.4**).

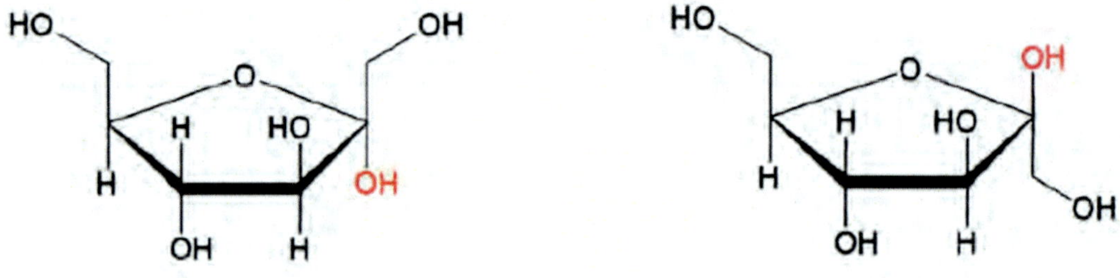

α–D-fructofuranose β–D-fructofuranose

Fig. (1.4). Anomers of sugar fructofuranose.

CONCLUSION

Carbohydrate is a primary macronutrient in dietary component for humans. Depending upon the type of carbohydrate, each can show varying digestion properties. Due to various applications in living organisms, it is essential to understand the type of carbohydrate qualitatively and quantitatively as well.

VIVA VOCE

1. What is the classification of carbohydrates?

Ans. Carbohydrates are classified as mono-, oligo-, and polysaccharides.

2. Explain the difference between reducing sugars and non-reducing sugars?

Ans. Reducing sugars have free aldehyde or ketone groups whereas, non-reducing sugars do not.

3. What are anomers and epimers?

Ans. Anomers are epimers that differ from each other in the configuration of C1 (aldose) or C2 (ketose). And epimers are stereoisomers that differ by one carbon atom only.

REFERENCES

[1] Nelson, D.L.; Lehninger, A.L.; Cox, M.M. *Lehninger principles of biochemistry*; Macmillan, **2008**.

[2] Berg, J.M.; Tymoczko, J.L.; Stryer, L. *Biochemistry (Loose-Leaf)*; Macmillan, **2007**.

[3] Jain, J.L. *Fundamentals of biochemistry. S*; Chand Publishing, **2004**.

[4] Bartnik, M; Facey, PC *Glycosides. In Pharmacognosy*; Academic Press, **2017**, pp. 101-161.

CHAPTER 2

Qualitative Tests for Carbohydrate Detection

Mrittika Sarkar[1], **Sinchan Hait**[2,*] and **Sai Joshi**[3]

[1] *Biomedical Engineering, University of Michigan, Michigan, USA*

[2] *Department of Chemical Engineering, BITS-Pilani K.K. Birla Goa Campus, Goa 403726, India*

[3] *CSIR-NCL, Dr. Homi Bhabha Road, Pune - 411 008, India*

Abstract: Qualitative tests for the determination of carbohydrates in the given sample are a crucial part of before getting into major analytical procedures. Every mono-, oligo-, and poly-saccharide vary depending upon the number of carbohydrate molecule present in it as well as changes in the side chains. Thus, every carbohydrate molecule has distinct functions and properties. Depending upon these physio-chemical properties of the carbohydrates, they respond to certain specific chemical reactions under certain conditions. Only mono- and dis-saccharides respond to the solubility test as they are soluble in water at room temperature. The Molisch test is only for the determination of the presence of carbohydrate, not depending on the types of it. The iodine test gives a result for polysaccharides. Whereas Fehling, Benedict and Osazone tests distinguish between reducing and non-reducing sugars. The Bradford test differentiates between mono- and di-saccharide-reducing sugars. Seliwanoff test is only for sucrose which is a non-reducing sugar. Bial's test is to determine the presence of pentose sugars.

Keywords: Bradford test, Bial's test, Benedict test, Carbohydrate, Fehling test, Iodine test, Monosaccharide, Molisch test, Oligosaccharide, Osazone test, Polysaccharide, Qualitative, Solubility, Seliwanoff test.

INTRODUCTION

To identify the presence of carbohydrates, if present or not in an unknown sample, a certain standardised preliminary protocol is to be followed. There have been various ways to qualitatively determine of carbohydrate present in a given sample. These tests are based upon the physico-chemical properties of the carbohydrates: mono-, oligo-, and polysaccharide; and reducing and non-reducing sugars can be differentiated. Detection and characterizationof carbohydrates present in an unknown sample solution based on various qualitative chemical assays are discussed below (Fig. **2.1**).

* **Corresponding author Sinchan Hait:** Department of Chemical Engineering, BITS-Pilani K.K. Birla Goa Campus, Goa 403726, India; E-mail: sinchan.hait1@gmail.com

Solubility Test

Aim

To determine the type of carbohydrates present in each sample by solubility test.

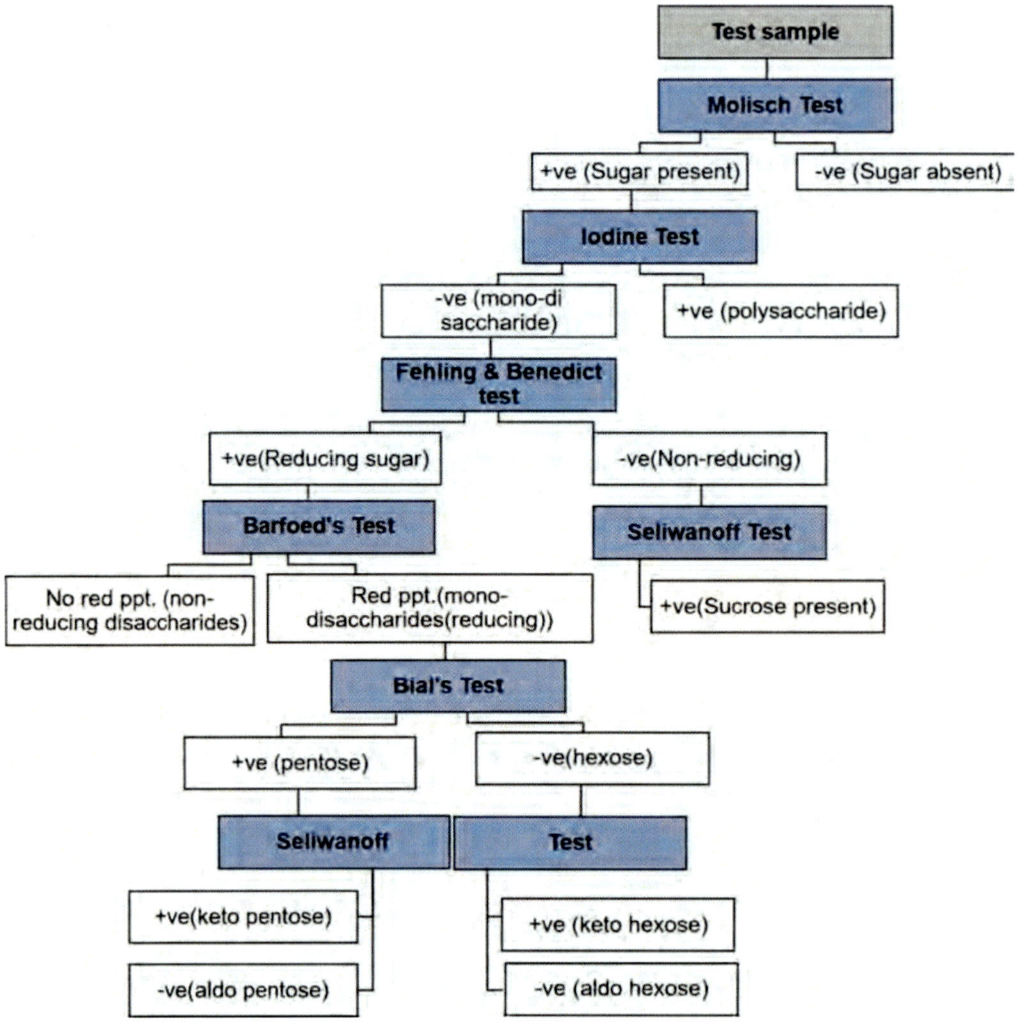

Fig. (2.1). Summary of all qualitative tests for carbohydrates.

Principle

Both monosaccharides (*e.g.*, glucose) and disaccharides (*e.g.*, sucrose) are crystalline solids at room temperature, but they are soluble in water because each molecule has several -OH groups that readily engage in hydrogen bonding with water molecules. Whereas in the case of polysaccharides, they have a complex structure which has strong interaction among polysaccharide molecules via hydrogen bonds, so they are insoluble in water [1, 3].

Requirements

❖ ***Glasswares***

• Test tubes

❖ ***Other Requirements***

• Dropper

• Test tube holder

❖ ***Reagents***

• Sugar solution (5% glucose, 5% sucrose, 1% starch)

• Distilled water

Procedure

To 1ml of sugar solution, add a few drops of distilled water and mix well.

Applications

• This test can help to determine whether mono-disaccharides or polysaccharides are present in the solution and their ability to dissolve in water.

Conclusion

If it solubilizes, then it confirms the presence of monosaccharides or disaccharides (glucose, sucrose) or else the presence of polysaccharides (starch).

Molisch's Test

Aim

To determine whether carbohydrates are present in a given sample by the Molisch test.

Principle

The Molisch test is a qualitative test to detect carbohydrates. Mainly monosaccharides, disaccharides, polysaccharides and glycoproteins (biomolecule complex of protein and sugar) can be determined using this test. The test has been named after Czech- Austrian botanist Hans Molisch.

When monosaccharide is treated with conc. H_2SO_4 or conc. HCl, OH groups of sugar are removed in the form of water, after which furfural is formed from pentose sugar and hydroxymethyl furfural from hexose sugar (Fig. **2.2**). These products (furfural and hydroxymethylfurfural) react with sulphonated α- naphthol to give a purple or violet-coloured complex at the interface of the acid and test layer (Fig. **2.2**) [1, 2]. If the test sample contains a poly- or disaccharide, a glycoprotein or glycolipid, the acid will first hydrolyse them into monosaccharide components, after which they get dehydrated to form furfural or its derivatives.

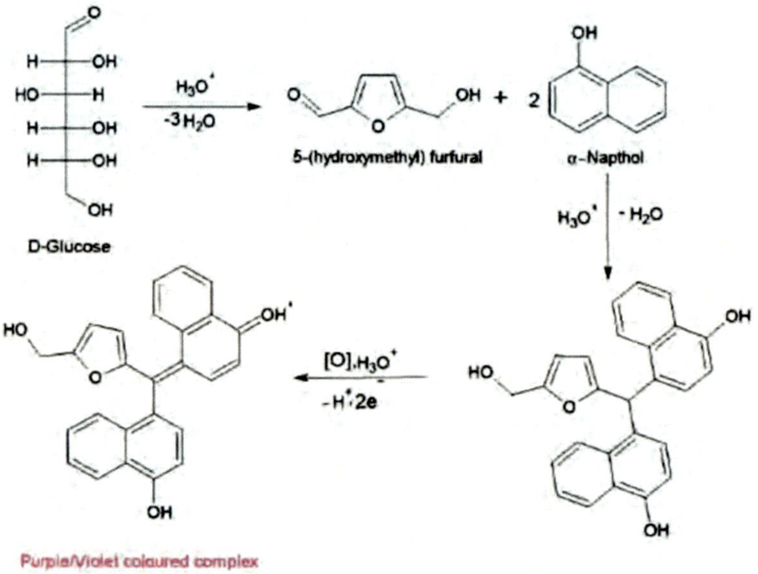

Fig. (2.2). Formation of purple or violet-coloured complex.

Requirements

❖ <u>*Chemicals*</u>

• Test solution (5% glucose solution)

• Molisch's reagent (Add 5% (w/v) α-naphthol in 95% ethanol or dissolve 500 mg α-naphthol in 10 ml 95% ethanol)

- Conc. H_2SO_4

❖ ***Glasswares***

- Test tubes
- Pipette (1 ml Beral pipette)

❖ ***Other Requirements***

- Test tube holder
- Dropper
- Distilled water

Procedure

1. Take 2ml of test solution in a dry test tube.
2. As a control take 2ml of distilled water in another tube.
3. Add 1-2 drops of Molisch's reagent to both tubes.
4. Gently pipette 1ml conc. H_2SO_4 and pour it along the side of the tube to get two distinct layers.
5. Observe the change in colour at the junction of two layers.

Conclusion

A purple colour between the junction of 2 layers indicates the presence of carbohydrates in the sample. No appearance of the purple complex indicates a negative test. However, if the appearance of green ring observation occurs, then it indicates that other impurities are present in the reagent, which interacts with α-naphthol and the conc. acid. The presence of concentrated sugar solution in the test sample yields a red ring due to partial charring of the sugar by the acid.

Precautions

- Do not pour conc. sulfuric acid directly into the solution as it can cause charring of carbohydrates due to heat generated during the reaction, which will result in a black ring, giving a false negative test.
- The α-naphthol solution should be prepared fresh as it is unstable.

Applications

This test can be used most practically to detect the presence of sugar in food products that claim to be sugar-free.

Fehling Test

Aim

To determine whether reducing sugars are present in a given sample by Fehling's test.

Principle

This test is used to differentiate between water-soluble carbohydrate and ketone functional groups and can be used supplementary to the Tollens' reagent test. It was developed by German chemist Hermann von Fehling.

Fehling's test is one of the sensitive tests for the detection of reducing sugars. Fehling's test is given positive by reducing sugars containing specifically aldehyde and alpha-hydroxy ketone groups. Fehling's reagents consist of two solutions: Fehling's solution A, which is aqueous copper sulphate and Fehling's solution B, sodium potassium tartrate (Rochelle salt), a strong alkali. This Fehling's B solution present in the reagent acts as the chelating agent in this reaction. Both Fehling A and B solutions are mixed in equal amounts to make the Fehling reagent.

On heating the sample with Fehling's reagent, aldehydes oxidize, giving a positive result, but ketones do not unless they are α-hydroxy ketones. The bis-tartrato cuprate(II) complex oxidizes the aldehyde to a carboxylate anion or aldonic acids, and in the process, the copper(II) ions of the complex are reduced to copper(I) ions. Red copper(I) oxide then precipitates out of the reaction mixture, which indicates a positive result (Fig. **2.3**) [1, 2]. The ketoses are oxidized to yield shorter chain acids.

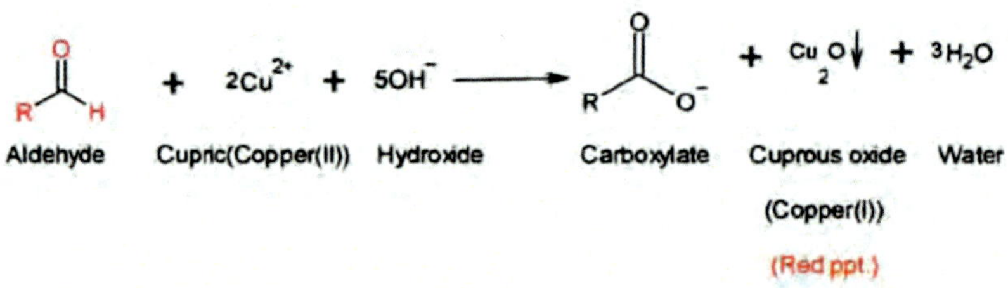

Fig. (2.3). Formation of cuprous oxide (red precipitate).

Requirements

❖ *Chemicals*

- Test solution (5% glucose or 5% sucrose)

- Fehling reagent (Fehling's "A": Dissolve 7 g $CuSO_4$. $5H_2O$ in distilled water containing 2 drops of dilute sulfuric acid. Fehling's "B": Add 35g of potassium tartrate and 12g of NaOH in 100 ml of distilled water. Mix both A and B in equal quantities to make Fehling's reagent).

❖ *Glasswares*

- Test tubes

- Pipette (1 ml Beral pipette)

❖ *Other Requirements*

- Test tube holder

- Water Bath (100 °C)

Procedure

1. Take a test tube and add 5 ml of the test solution to it.

2. Add Fehling's reagent to this test tube.

3. Shake the test tubes properly and gently place them into boiling water for a water bath.

4. Carefully observe the test tubes and note down the result of each test tube after 5 minutes.

Conclusion

The appearance of a reddish-brown precipitate indicates a positive result, thus the presence of reducing sugar (*e.g.*, Glucose, fructose, lactose). And the appearance of blue colour and no precipitate is the presence of non-reducing sugars (*e.g.*, sucrose, starch).

Precautions

- Fehling's solution is corrosive and toxic, wear appropriate personal protective gear while performing this experiment.

Applications

- It is used to screen for glucose in the urine, thus detecting diabetes.

- This test is used in the breakdown of starch to glucose syrup and maltodextrins, to measure the amount of reducing sugar, thus revealing the dextrose equivalent (DE) of the starch sugars.

Benedict's Test

Aim

To determine whether reducing sugars are present in a given sample by Benedict's test.

Principle

There are generally two types of sugar: reducing (all monosaccharides, as they have free anomeric carbon in their structure, *i.e*., free aldehyde or ketone group that can reduce cupric salt) and non-reducing (some disaccharides and polysaccharides). Benedict's reagent is sensitive and can be used to determine and demonstrate the reducing property of carbohydrates (even to small quantities of reducing sugars (0.1%)) and was discovered by Stanley Rossiter Benedict, an American chemist [2].

When reducing sugar is heated with Benedict's reagent, due to the presence of the alkaline sodium carbonate, the reducing sugar gets converted into a strong reducing agent called enediol, which decreases the cupric particles (Cu^{2+}) present in the reagent to cuprous particles (Cu^{+}), which appear as insoluble red copper oxide (Cu_2O) (Fig. **2.4**) [1, 2]. This is a semi-quantitative test, as the colour of the precipitate can indicate the approximate quantity of sugar present in the sample.

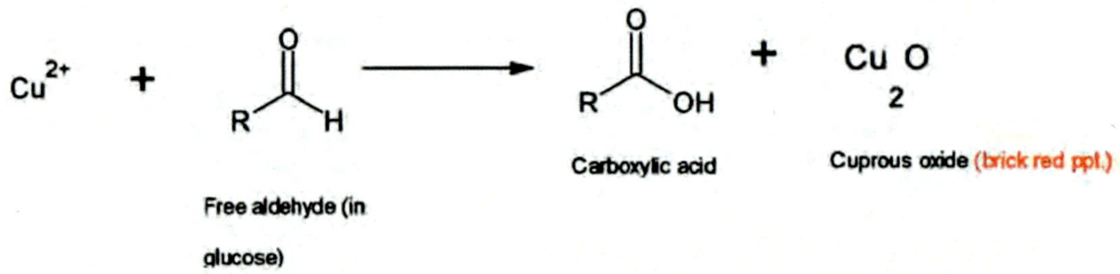

Fig. (2.4). Formation of cuprous oxide (brick red precipitate).

Requirements

❖ ***Chemicals***

• Sample solution (5% glucose or 5% sucrose)

• Benedict's reagent (A deep-blue alkaline chemical reagent which consists of copper sulphate pentahydrate ($CuSO_4.5H_2O$) in the Citrate buffer. Dissolve 17.3 g sodium citrate and 10 g sodium carbonate in 85 ml H_2O. Filter it. Dissolve 1.73 g $CuSO_4.5H_2O$ in 10 ml H_2O, followed by the addition of carbonate-citrate mixture).

❖ ***Glasswares***

• Test tubes

• Pipette

❖ ***Other Requirements***

• Test tube stand and holder

• Hot water Bath

• Bunsen burner

Procedure

1. Pipette out 2 ml (10 drops) of Benedict's reagent and place it in the clean test tube.

2. Add approximately 1 ml of sample to Benedict's reagent.

3. Water bath this sample for 3-5 minutes (can be heated directly over a flame).

4. Observe for precipitate formation in the test tube.

Conclusion

A reddish precipitate within three minutes indicates the presence of reducing sugars present. Example: Glucose. No precipitation indicates the absence of reducing sugar and probably non-reducing sugar. Example: Sucrose.

Precautions

1. Benedict's solution is an irritant, so avoid contact with the skin and eyes.

Applications

Benedict's test can be used to detect the presence of glucose (reducing monosaccharide) in urine.

Seliwanoff's Test

Aim

To detect the presence of aldose and ketose sugar by Seliwanoff's test.

Principle

Seliwanoff's test is used to differentiate between aldose and ketose sugars, and it is specific for ketohexoses. This test is very similar to Bial's test.

When the sample is heated with concentrated acid, samples with ketoses are dehydrated more rapidly to give furfural derivatives, and condensation with resorcinol gives cherry red complex (Fig. **2.5**) [1, 2]. If the reaction is allowed for a longer time (more than 10 minutes) for samples with aldoses, it may also produce positive results as the reaction with aldehydes takes place slowly. Seliwanoff's test can distinguish between two monosaccharides, *e.g.* glucose (aldehyde group present) and fructose (ketone group present).

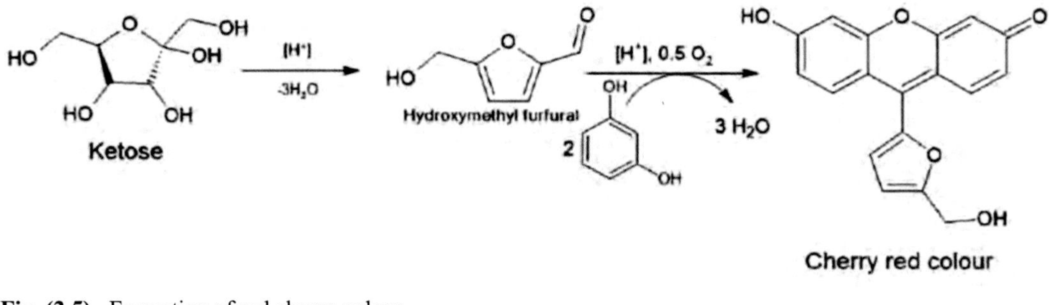

Fig. (2.5). Formation of red cherry colour.

Requirements

❖ ***Chemicals***

• Test solution (5% fructose or 2% glucose)

• Seliwanoff's reagent (Dissolve 110 mg of resorcinol in 220 ml of 3N HCl or dissolve 50 mg resorcinol in 33 ml conc. HCl, then make it up to 100 ml with H_2O).

❖ ***Glassware***

• Pipettes

• Test tubes

❖ ***Other Requirements***

• Water bath

• Test tube holder

Procedure

1. Add 5 ml of Seliwanoff's reagent and 1 ml of a sample in a test tube.

2. Heat the solution using a water bath in boiling water for 1-5 minutes and observe for colour change.

Conclusion

A positive test is indicated by the colour of the sample turning to cherry red immediately, which means keto sugar is present in the solution (*e.g.*, fructose). If the red cherry complex is not formed immediately, then wait for about 10 minutes; if a slower framing pink colour is seen, then aldehydes are present, indicating a positive test (*e.g.*, glucose). Otherwise, it is a negative test (No colour is seen.)

Precautions

• Make sure to add 5 ml of reagent and 1 ml of material to be tested because this is for the exact compatibility of both agents.

• Make sure that what's inside the test tube never contacts anyone because it is highly corrosive and can bring serious damage to the skin.

• Do not boil glucose for long as prolonged boiling can transform it to fructose by the catalytic action of acid and form a cherry red complex.

Troubleshooting

• Glucose or other sugar in high concentration may interfere by producing similar coloured compounds with the Seliwanoff reagent. So, the amount of glucose for this test must not be present more than 2%.

Applications

• Seliwanoff's colour reaction is used in the method for the colourimetric determination of fructose in fermentation media.

• This test can also determine the presence of sucrose and inulin as these are hydrolyzed by acid to give fructose.

Bial's Test

Aim

To determine whether pentose sugar is present in a given sample by Bial's test.

Principle

The Bial's reagent contains HCl that dehydrates pentoses to form furfural and hexoses to form 5- hydroxymethyl furfural. After that, furfural reacts with orcinol and ferric chloride ($FeCl_3$) to produce a blue-green complex, while 5-hydroxymethyl furfural produces a muddy brown colour complex see Fig. (**2.6**). This colour formation indicates the presence of pentose sugar [1, 2].

Requirements

❖ ***Chemicals***

• Test sample (5% ribose or 5% glucose solution)

• Bial's Reagent (Dissolve 300 mg of orcinol in 5 ml ethanol. Now, in 100 ml of 0.1% solution of $FeCl_3.6H_2O$, add 3.5 ml of this mixture).

❖ ***Glassware***

• Test tubes

• Pipettes

❖ ***Other Requirements***

• Test tube holder and stand

• Bunsen burner

• Water bath

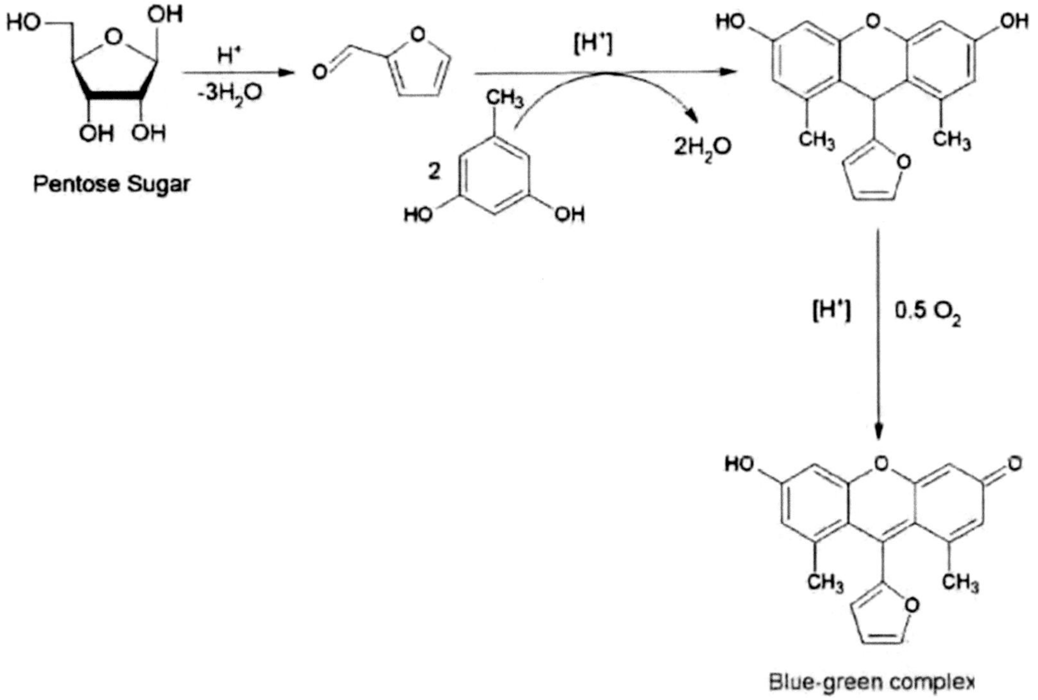

Fig. (2.6). Formation of Blue-green complex.

Procedure

1. In a clean, dry test tube, add 1 ml of test solution.

2. Add 2 ml of Bial's reagent to this tube and mix well.

3. Heat the test tube on a Bunsen burner gently or keep the tube in a boiling water bath for one minute and then cool it down to room temperature.

4. Observe the appearance of colour.

Conclusion

A positive test is indicated by the formation of bluish green colour which indicates pentoses. A negative test indicates the presence of hexose sugar, which generally gives green, brown or red products.

Precautions

1. The colour formed in the test results depends upon the quality of the reagents. So care should be taken while preparing the reagents, and they should be prepared

fresh. If commercially available reagents are used, assure that it is not kept open for a long time.

2. Do not rapidly cool the sample after heating, otherwise, the results may be affected.

3. Do not boil the solution for a long time as it may give a similar blue-green colour due to the formation of gluconates in the sample.

Application

This test is used to differentiate pentose sugar from hexose sugar.

Iodine Test

Aim

To test the presence of starch by Iodine test.

Principle

The iodine test is used to distinguish mono- or di-saccharides from certain poly-saccharides. This test is best to identify starch. Starch is a polysaccharide found in plants and consists of two different types of polysaccharides, which are connected in two different ways: linear amylose (10-20%) and amylopectin (80-90%) [4].

Molecular iodine (I_2) is not easily soluble in water, which is why potassium-iodide is added. This combination makes up the aqueous solution called Lugol's Iodine. Together, they form polyiodide ions, for example, I_3 and I_5 [4].

It is observed that the coil or spring structure of the glucose chain is the key to this test and the colour depends on the length of the glucose chains.

The I_3 (triiodide) and I_5 (pentaiodide) ions formed are linear and slip inside the helix structure. The transfer of charge between the glucose chains and polyiodide ions results in changes in the spacing of the energy levels, which absorbs visible light, giving the complex colour [4].

The colours are caused by so-called charge transfer (CT) complexes. For *e.g.*, amylose in starch is the compound that is responsible for the blue colour. Its chain forms a helix shape (and acts as a charge donor), and polyiodide (as a charge acceptor) binds inside this helix. This forms the CT complex, which absorbs light and turns dark blue (Fig. **2.7**) [4].

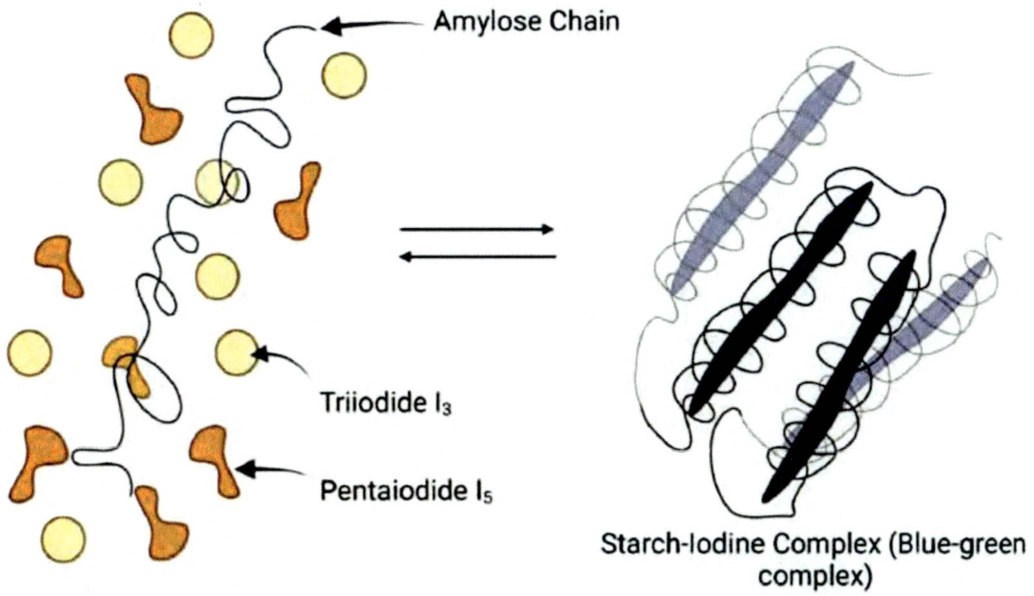

Fig. (2.7). Formation of Charge-Transfer (CT) complex.

Requirements

❖ ***Chemicals***

• Starch sample

• Iodine solution/Lugol's iodine (10 g of KI is dissolved in 100 ml water and then add 5 g I_2 crystals to this solution and mix. Store in a brown bottle)

❖ ***Glasswares***

• Test tubes

❖ ***Other Requirements***

• Test-tube stand and holder

• Dropper

Procedure

1. Take the sample in a test tube.

2. Add 2-3 drops of Lugol's iodine with the help of a dropper.

3. Observe the colour change.

Conclusion

The positive test indicated by the colour of the sample changes to dark blue-black. Where a negative test is indicated by the no colour formation in the sample. It will remain orange-brown or yellow (Lugol's Iodine colour).

Precautions

Do not use too much iodine solution.

Troubleshooting

• If an unknown sample, instead of a blue-black colour gives purple, reddish-brown, and orange-yellow hue colour, then residues of dextrin, glycogen, and amylopectin, respectively, are present in the sample.

Application

1. This test can be used for the assessment of hyperhidrosis in amputees.

2. It can be used as a diagnostic tool for Horner Syndrome.

3. This test is used to detect the presence of starch in various samples.

4. It is used to test the process of photosynthesis in plants by looking for the presence of starch.

Barfoed's Test

Aim

To detect the presence of monosaccharides in a sample by Barfoed's test.

Principle

Barfoed's test is used to detect reducing monosaccharides in an unknown sample. This test can also be used for disaccharides, but the only drawback is that the reaction takes place very slowly.

Barfoed's reagent consists of ethanoic (acetic) acid and copper (II) acetate, which, when combined with the sample solution and boiled, gives a red copper (II) oxide precipitate, which indicates the presence of reducing sugar [1, 2]. Acidic pH is unfavourable for reduction, so monosaccharides, which are strong reducing agents, react in about 1-2 min while reducing disaccharides takes a longer time of about 7-8 minutes, having first to get hydrolysed in the acidic solution and then react with the reagent. The positive result is indicated by a thin red precipitate formation. The difference in the time of appearance of precipitate thus helps distinguish reducing monosaccharides from reducing disaccharides (Fig. **2.8**).

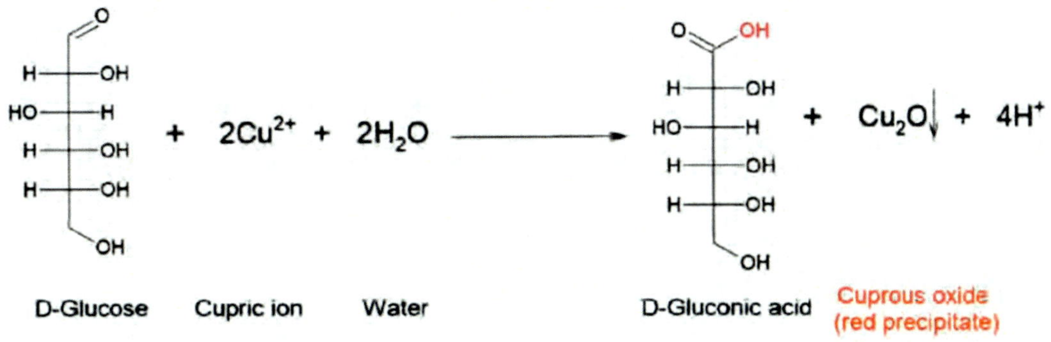

Fig. (2.8). Formation of D-gluconic acid and cuprous oxide (red precipitate).

Requirements

❖ *Chemicals*

- Test solution (5% glucose)

- Barfoed's reagent (Add 0.33M solution of copper acetate to 1% acetic acid. Freshly prepared reagent should be used).

❖ *Glasswares*

- Test tubes

- Pipettes

❖ *Other Requirements*

- Distilled water

- Test tube holder and stand

- Water bath

- Vortex

Procedure

1. Take 1 ml of the sample in a clean, dry test tube and 1 ml of distilled water as a control in another tube.

2. Add 2-3 drops of Barfoed's reagent in both tubes and mix them using a vortex.

3. Keep both test tubes in the water bath for 1-2 minutes, but it should not exceed more than 2 minutes as the disaccharides might be hydrolyzed into monosaccharides and give a positive result.

4. Observe the appearance of colour in the test tubes and note down the time taken for the colour to appear.

Conclusion

A positive test is indicated by the formation of red precipitate and detects the presence of reducing monosaccharides in the sample if the colour appears within the first few minutes and if the colour appears later than the first 3 minutes, the sample has reducing disaccharides. A negative test can be identified by the absence of red precipitate.

Precautions

1. The test solution (if it is a disaccharide) should be made to 1%(w/v) because if a higher concentration of disaccharides is present in a sample, it might give a positive result.

2. Barfoed's reagent should be added in more quantity than the original sugar solution as this ensures a good reaction and gives a positive result.

3. Tracking of time should be done very carefully, and it should be ensured that the test tubes are not boiled/water bath for more than 2 minutes.

Application

• This test is used to identify and distinguish reducing monosaccharides from reducing disaccharides.

Osazone Test

Aim

To detect the presence of reducing sugars in the sample solution by Osazone Test.

Principle

The osazone test is used to detect reducing sugars based on the time of appearance of the complex and was developed and used by Emil Fischer. This test is also termed the Phenylhydrazine test because the reagent of this test consists of phenylhydrazine in acetate buffer. This test is based on the fact that carbohydrates with free carbonyl groups react with phenylhydrazine to form osazone.

The condensation-oxidation-condensation reaction between three molecules of phenylhydrazine and carbon one and two of aldoses and ketoses yields 1, 2-diphenylhydrazine, which is known as osazone and can be seen as yellow-coloured crystals (Fig. **2.9**) [5]. These crystals have characteristic shapes, solubility, and melting points, and their time of formation differs for different sugars.

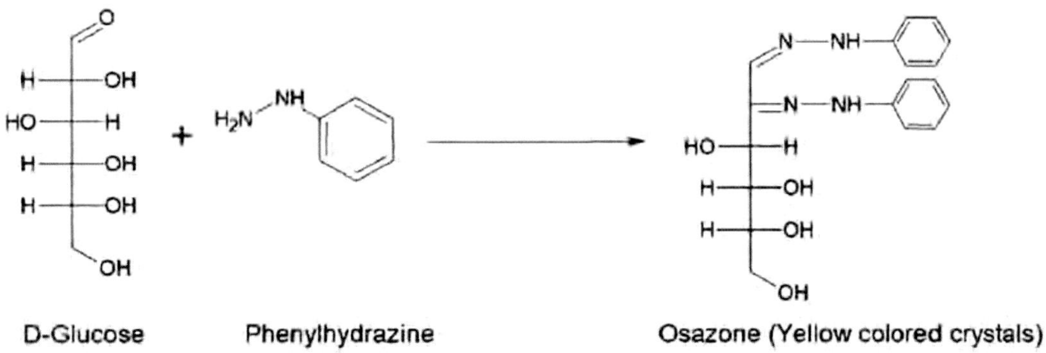

Fig. (2.9). Formation of 1,2-diphenylhydrazine.

When a molecule of phenylhydrazine hydrochloride comes in contact with reducing sugar, it reacts to form phenylhydrazine hydrochloride. This complex reacts once again with 2 more molecules of phenylhydrazine hydrochloride (to 1st give an intermediate keto derivative and then osazone). Additional phenyl hydrazine is consumed during the reaction with monosaccharides, to oxidize the adjacent OH-group to the carbonyl group which then forms a second phenyl hydrazone. Such bisphenyl hydrazones are called osazones [5].

Requirements

❖ *Chemicals*

- Osazone mixture (Add 0.5 g of phenylhydrazine hydrochloride and 0.1 g of sodium acetate)

• Glacial acetic acid

❖ ***Glasswars***

• Test tubes

• Pipettes

❖ ***Other Requirements***

• Test tube stand

• Water bath

• Dropper

• Vortex

• Microscope

Procedure

1. In a clean, dry test tube, take 5 ml of test solution.

2. Add 1 ml of osazone mixture and 2 drops of glacial acetic acid to this test tube.

3. Mix it well using a vortex and heat the mixture under a b oiling water bath maximum for 30 minutes till a ppt. forms (a ppt. formation indicates that we can remove the sample from the water bath).

4. Cool the solution slowly and observe the formation of crystals at various time points.

5. Observe the shape of the crystal under low magnification under a microscope to identify the sugar in the sample.

Conclusion

The formation of yellow-coloured crystals will be seen upon cooling, which means the presence of reducing sugar. Different sugars can be identified based on their time of appearance, shape and structure see Table **2.1**. No yellow colour crystal formation occurs upon cooling, a negative result.

Table 2.1. Characteristics of each sugar with Osazone test.

Name of Sugar	Shape of Crystal	Time for Crystal Formation (Min)
Lactose	Cotton-ball	30-45
Maltose	Sunflower/flower shape	30-45
Xylose	Fine-long needle	15
Arabinose	Balls with needles	15
Galactose	Balls with thorny edges	10-20
Mannose	Needle	10
Fructose	Needle	2-4
Glucose	Needle	5-6

Precautions

1. After a water bath, cool the sample gradually to room temperature. Do not cool under running tap water or ice otherwise, crystal formation will not take place properly.

2. Do not add excess glacial acetic acid otherwise, the pH of the reaction will reduce, resulting in the hydrolysis of disaccharides to monosaccharides, and osazones formed out of this reaction will be insoluble in a hot solution.

Troubleshooting

- If there is an appearance of red colour in the solution after the heating process, it means that the solution has concentrated due to the boiling process, and no crystals will separate in the concentrated form. So, diluting with water is necessary for the separation of crystals.

Application

- Used to identify different types of reducing sugar.

- Different characteristic crystal formations of osazones with each sugar can be used to identify and differentiate sugars in biological fluids like urine.

- Used for locating sugars in plant tissues.

- This test can be used as a simple chemical test for glucose in urine in poorly controlled diabetes mellitus.

DID YOU KNOW?

1. The Molisch test is not only specific for carbohydrates but also other components like furfurals, furfural-yielding substances, and organic acids, *e.g.*, citric acid, lactic acid, oxalic acid, formic acid, *etc.*

2. A counterfeit banknote detection pen uses an iodine test to detect counterfeit banknotes printed on starch-containing paper. Banknotes are made of cotton paper or sometimes mixed with linen, abaca or other types of fibres to increase their life. When a test solution containing iodine is applied to a test area of a banknote, the counterfeit banknote turns bluish-black or black if they are made of normal paper which contains starch, whereas, in the case of a genuine banknote, the test area is stained golden-brown.

3. *Camponotus saundersi*, an ant species found in Malaysia and Brunei, sprays a sticky secretion (toxic glue) in defence which also consists of orcinol. This glue has corrosive properties which function as a chemical irritant and can entangle and immobilize all nearby victims. The irritant is responsible because of the presence of orcinol.

4. Carbohydrates (sucrose) play a role in blooming in rose (*Rosa hybrida L.*) flowers. Levels of non-reducing sugars and sucrose were seen to increase gradually with flower bud development between 2003-05 in an experiment conducted on two cultivars of cut-rose (*Rosa hybrida L.*), 'Grandgala' and 'First Red'.

CONCLUSION

In this chapter, several qualitative tests are discussed. Based on the chemical properties of the carbohydrates, they react with certain chemical components and produce some colour compounds, which then can be interpreted as the presence of carbohydrates in the sample. However, these techniques can be used for primary analysis purposes only. They provide an idea of the presence of carbohydrates or, to some extent, types of carbohydrates. Further, quantifying carbohydrates needs to be done, which is discussed in Chapter 3.

VIVA VOCE

1. Which carbohydrates are soluble in water - Monosaccharides or Disaccharides?

Ans. Both are soluble as they have several OH groups that readily engage in hydrogen bonding with water molecules.

2. How do monosaccharides give a positive test in the Molisch test?

Ans. When monosaccharide is treated with conc. H_2SO_4 or conc. HCl, OH groups of this monosaccharide are removed in the form of water, after which furfural is formed from pentose sugar and hydroxymethyl furfural from hexose sugar. These products react with sulfonated α- naphthol of Molisch reagent to give a positive test.

3. Which type of sugar gives a positive Fehling's test and because of which functional groups?

Ans. Positive Fehling's test is given by reducing sugars containing specifically aldehyde and alpha hydroxy ketone groups.

4. How to check whether a sample has given a positive Benedict's test or not? If the solution gives a negative result, then what is the colour change that indicates it?

Ans. If there is a formation of a reddish precipitate within three minutes of water bath/boiling, then the solution has given us a positive Benedict's test.

The negative test is given a blue colour (as the colour of the reagent is blue).

5. If you are given glucose and a fructose solution, which one of the two will give you a positive Seliwanoff test and why?

Ans. Both the solutions will give a positive solution as Seliwanoff's test is used to detect the presence of aldose (glucose) and ketose (fructose) sugars. The fructose solution, after being heated with concentrated acid, dehydrates quickly to give a cherry red complex indicating a positive test, whereas the glucose solution gives a positive test (pink colour) only when the solution after being heated with the conc. acid is allowed to remain for a longer time (10 min or more).

6. Bial's reagent contains HCl, orcinol and ferric acid as its components. Out of these 3 components, which of them acts as a dehydrating reagent and condensation reagent?

Ans. HCl acts as a dehydrating acid; orcinol and ferric chloride act as condensation reagents.

7. If a sample gives purple or reddish-brown colour as its result in the Iodine test, what can we infer from this?

Ans. It can be inferred that residues of dextrin and glycogen are present in the sample, which gives purple and reddish-brown colour, respectively.

8. Why do reducing disaccharides take longer to show positive results than reducing monosaccharides in Barfoed's test?

Ans. Reducing disaccharides take a longer time because they have to first get hydrolysed in the acidic solution and then react with the reagent to give the result, unlike monosaccharides.

9. Which qualitative test for carbohydrates is also known as the "Phenylhydrazine test"? What characteristic shape of crystal does maltose sugar show under a microscope in this test?

Ans. The osazone test for carbohydrates is also termed the Phenylhydrazine test. Maltose shows a flower or sunflower shape under a microscope.

10. How does monosaccharide give a positive test in the Molisch test?

Ans. When monosaccharide is treated with conc. H_2SO_4 or conc. HCl, OH groups of this monosaccharide are removed in the form of water after which furfural is formed from pentose sugar and hydroxymethyl furfural from hexose sugar. These products react with sulphonated α- naphthol of Molisch reagent to give a positive test.

REFERENCES

[1] Tiwari, A. *Practical Biochemistry: A Student Companion*; LAP LAMBERT Academic Publishing, **2015**, pp. 4-10.

[2] G.D., K. *Practical Biochemistry*; Jaypee Brothers Medical Publishers, **2011**, pp. 3-15.

[3] Guo, M. Q.; Hu, X.; Wang, C.; Ai, L. Polysaccharides: Structure and solubility. *Solubility of polysaccharides,* **2017**, 7-21.

[4] Ribeiro, L.; Rocha, R.; Martins, J.; Monteiro, A. Starch–iodine test: A diagnostic tool for horner syndrome. *BMJ Case Rep.,* **2020**, *13*(9), e238541. [http://dx.doi.org/10.1136/bcr-2020-238541] [PMID: 32913069]

[5] Md, D.T.S.; Modi, N. Utility of osazone test to identify sugars. *J. Med. Sci. cli. Res.,* **2016**, *4*(12), 14361-14365. [http://dx.doi.org/10.18535/jmscr/v4i12.14]

CHAPTER 3

Quantification Tests for Carbohydrate Detection

Mrittika Sarkar[1], **Anshika Sah**[2], **Payal Agarwal**[3], **Sai Joshi**[4] and **Sinchan Hait**[5,*]

[1] *Biomedical Engineering, University of Michigan, Michigan, USA*

[2] *Environmental Studies and Resource Management, TERI School of Advanced Studies, New Delhi - 110 070, India*

[3] *Graduate School of Medical Science and Engineering, Korean Advanced Institute of Science and Technology, Daejoeon, South Korea*

[4] *CSIR-NCL, Dr. Homi Bhabha Road, Pune - 411 008, India*

[5] *Department of Chemical Engineering, BITS-Pilani K.K. Birla Goa Campus, Goa-403726, India*

Abstract: Quantification of carbohydrates by the anthrone method is a colourimetric assay. In this method, first, the complex carbohydrates are hydrolysed in a highly acidic medium, resulting in monosaccharides. Thereafter, these monosaccharides are dehydrated to form 5-hydroxymethylfurfural (5-HMF) or 2-*furfural* (2-F), followed by reacting with anthrone reagent. This reaction produces a blue-green complex which is colourimetrically determined. In DNSA method 3,5-dinitrosalicylic acid (DNSA), a yellow colour reagent reacts with reducing sugar's carbonyl group and forms orange-red colour compound 3-Amino-5-nitrosalicylic acid (ANSA). The quantity of carbohydrate present in a sample is determined by Beer-Lambert's law. Apart from these techniques Folin-Wu, Hexokinase, Phenol-Sulphuric acid, Somogvi-Nelson, and GOPOD method are discussed in this chapter in detail.

Keywords: Anthrone, Carbohydrate, Colourimetry, Folin-wu, GOPOD, Hexokinase, Phenol-sulphuric acid, Somogvi-Nelson, 5-hydroxymethylfurfural (5-HMF), 2-*furfural* (2-F), 3,5-dinitrosalicylic acid (DNSA), 3-Amino-5-nitrosalicylic acid (ANSA).

INTRODUCTION

Quantification of carbohydrates is an important step to determine the amount of carbohydrate present in a given sample. Several methods have been developed for the same. In this chapter, we will be dealing with colourimetric assays for the quantitative determination of carbohydrates. The assays include the Anthrone

* **Corresponding author Sinchan Hait:** Department of Chemical Engineering, BITS-Pilani K.K. Birla Goa Campus, Goa-403726, India; E-mail: sinchan.hait1@gmail.com

Pamela Jha (Ed.)

method, DNSA method, Folin-Wu method, Hexokinase method, Phenol-Sulphuric acid method, Somogvi-Nelson Method, and GOPOD method. In these assays, the carbohydrate present in the sample reacts with certain chemical ingredients under certain specified conditions and forms a colour complex which is then detected *via* spectrophotometer.

The colour formation is directly linked with the concentration of the carbohydrate present in the sample. This is based on Beer-Lambert's law which states that the absorbance of a solution is dependent upon the concentration of material present in the given solution and path length.

$$A = \varepsilon c l \tag{1}$$

Where, A=absorbance, ϵ=molar absorption coefficient ($Mol^{-1}cm^{-1}$), c=conentraion of the solution (M), l=path length (cm)

ANTHRONE METHOD

Principle

The anthrone method for carbohydrate analysis is used to quantify total carbohydrates in an unknown sample. The carbohydrate content can be measured by hydrolyzing the polysaccharides into simple sugars by acid hydrolysis and estimating the resultant monosaccharides, which are the basic units of all carbohydrates. This method then uses a colourimetric approach which requires the aid of a chemical agent (anthranol) to determine the concentration of a chemical compound in a solution [1 - 4]. This method takes place in three steps:

Hydrolysis of Complex Carbohydrates

In this step, complex carbohydrate polymers react with water to break down into simple units of monomers. The disaccharide or long chain of polysaccharides breaks into two unionized components with each water molecule splitting. Here one part gains a hydrogen atom (H+) and the other gains a hydroxyl group (OH–) from a split water molecule. However, this process can be catalyzed by carrying out the reaction in an acidic condition (much like how the human digestive system uses various enzymes lactase, sucrase, and amylase, collectively called glycosidase or invertase in honey bees). Catalysis of the cleavage of *β 1-4 glycosidic bonds*occurs via a nucleophilic substitution reaction where an electron-rich nucleophile (electrophilic carbon atom of aldehydes and ketones of carbohydrate) attacks a positively charged electrophile (from acid) to replace a leaving group [1 - 4]. The use of aqueous sulphuric acid (H_2SO_4) facilitates our purpose (Fig. **3.1**).

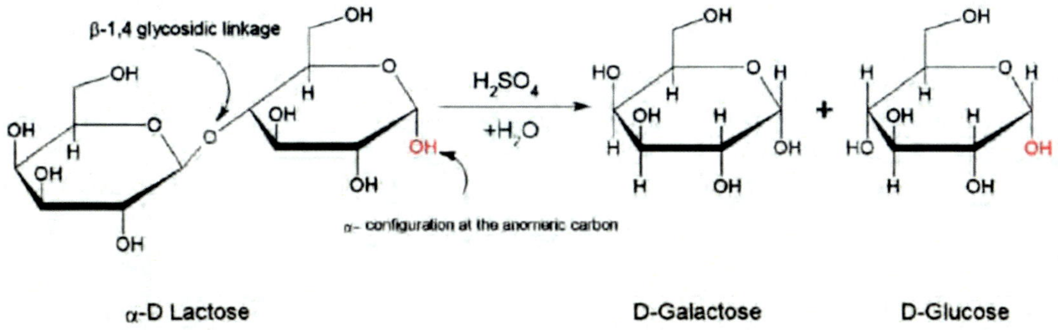

Fig. (3.1). Hydrolysis of carbohydrates.

Dehydration of Monosaccharide

Dehydration of 6-carbon monosaccharides or hexoses (*i.e.*, glucose, galactose, fructose) and 5-carbon monosaccharides or pentoses (*i.e.*, xylulose, arabinose, ribose) typically leads to the formation of 5-hydroxymethylfurfural (5-HMF) and 2-*furfural* (2-F) respectively [2, 3]. This step is observed in some qualitative tests for carbohydrates (Seliwanoff's test, Molisch's test) (Fig. **3.2**).

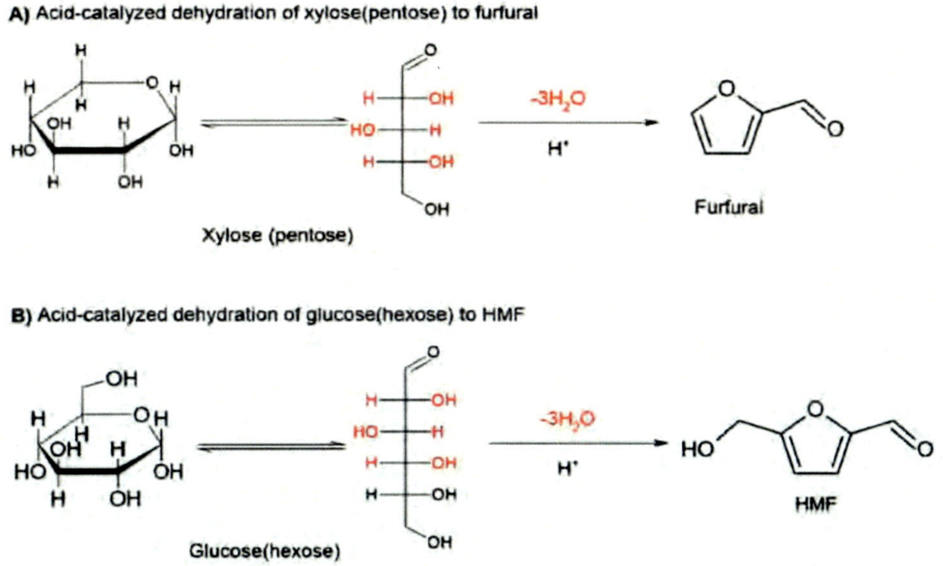

Fig. (3.2). Dehydration of monosaccharide.

Note - Ketose gets dehydrated faster than aldoses by different intermediates. Furfural is a furan aldehyde.

Reaction with Reagent

Anthranol is an active form of anthrone, which reacts with furfural and HMF to produce blue-green complexes. Hydrochloric acid or Sulphuric acid increases the rate of condensation and, thus, the formation of this coloured product. Formic acid reduces the degradation of the coloured complex while simultaneously inducing the colouration of the solution. The colour development is linear to the increasing quantity of the monosaccharide [4]. It gives a green colour to a solution low in carbs and a blue colour to more concentrated solutions (quantitative threshold). The intensity of the colour thus produced can be measured using a colourimeter at 620 nm (orange-red). Colourimetry analysis will give the absorbance measurement of the wavelength of light, which will show a rising trend with increased carbohydrate concentration (Fig. **3.3**)

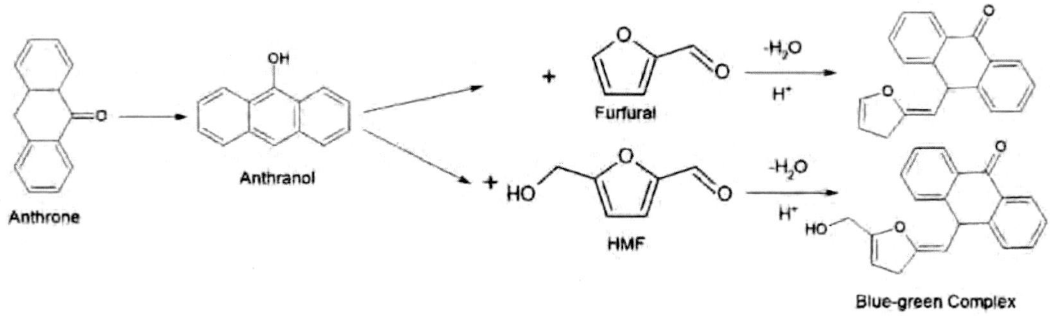

Fig. (3.3). Reaction with reagent.

Range of detection: 10 to 120 µg/mL for glucose.

Requirements

- Equipment - vortex mixer, mantle heater/water bath.
- Chemicals - anthrone reagent (See Appendix, conc. H_2SO_4, carbohydrate sample.
- Glucose stock solution: 200µg glucose per mL distilled water.
- Glassware- test tube, test-tube stand, pipettes, beaker, test tube caps, tissue paper, wash bottle.

Procedure

1. From a set of 8 test tubes: Consider the first tube as blank and label the rest as 1-7; tubes 1 to 6 will be used for the construction of a standard curve. Test tube 7 is for the unknown sample (Table **3.1**).

2. Pipette out into the test tubes 1 to 6 with different volumes of stock solution (200µg /ml) and make up the volume to 1mL with distilled water.

3. To each tube, add 5 mL of the anthrone reagent and mix well by a vortex.

4. Cover the top of the tubes with caps and incubate at 90°C for 17 minutes or boiling water bath for 10 minutes.

5. Cool to room temperature and measure absorbance at 620 nm against a blank (Table **3.1**).

6. Prepare a standard curve of absorbance *vs*. concentration of glucose see Fig. (**3.4**).

Table 3.1. Reference data for Anthrone method.

Sr. No.	Conc. of Glucose (µg /ml)	Stock Volume (ml)	Distilled Water (ml)	Anthrone Reagent (ml)	Incubate at 90°C for 17 Mins or in a Boiling Water Bath for 10 Mins	Absorbance (at 620 nm)
B	0	0	1	5	-	0
1	20	0.2	0.8	5		0.21
2	40	0.4	0.6	5		0.39
3	60	0.6	0.4	5		0.55
4	80	0.8	0.2	5		0.72
5	100	1	0	5		0.95
6	?	1ml (diluted 20 times)	-	5		0.41

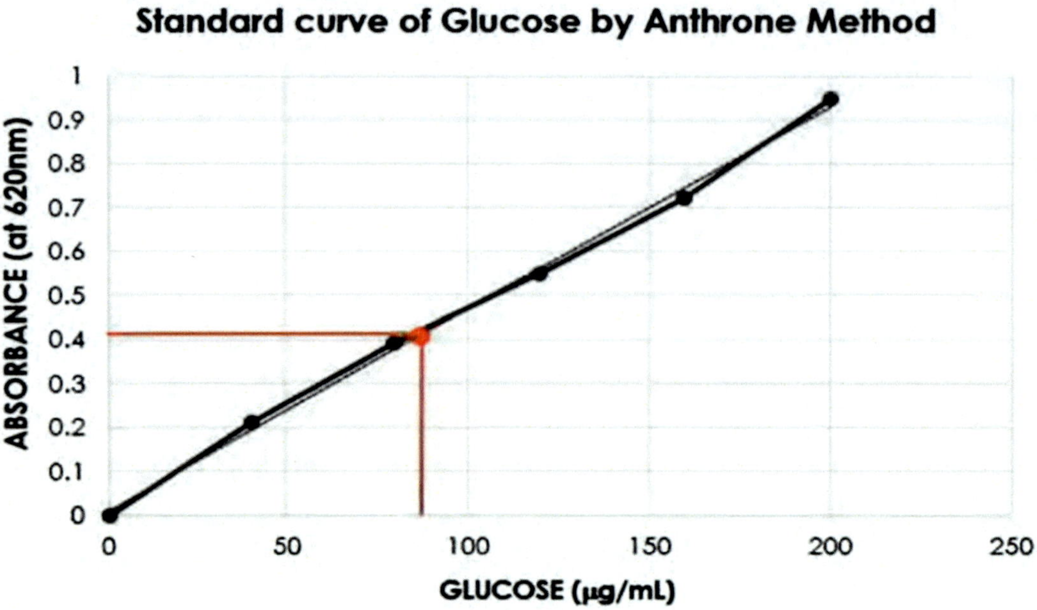

Fig. (3.4). Standard curve of Glucose by Anthrone Method. The standard graph provided is only for representational purpose. Data shown in this graph may differ.

Results

Example of a Glucose Standard Curve for Anthrone Method of Sugar Estimation.

Calculation:

1. Compute the linear equation for the absorbance obtained from the known samples.

2. Use the Eqn. 3 for calculating the concentration of the unknown sample.

$$\text{Linear equation: } y = mx + c \qquad (2)$$

$$Concentration\ of\ sample(\mu g/ml) = \frac{(OD\ of\ the\ Sample - c)}{m}\ X\ dilution\ factor \qquad (3)$$

Example:

The equation thus obtained: y = 0.0046x + 0.01

Concentration of unknown (diluted) (x) = 86.95 µg /ml

Concentration of unknown (dilution factor * x) = 20 * 86.95

= 1,739 μg /ml = 1.739 mg/mL = 173.9 mg/dL

Normal Blood Glucose Level: Fasting (**72 to 99 mg/dL**) and after eating (up to **140 mg/dL**)

(Hypoglycemic: further tests recommended to confirm the results)

Did you know?

• Grass-eating mammals like cows rely on microorganisms in their gut to convert cellulose to glucose.

• Bee converts nectar to honey by dehydration after breaking the sucrose in nectar into glucose and fructose.

• Freshly extracted honey displays HMF (5-hydroxymethylfurfural) levels lower than 5 mg/kg.

DNSA METHOD

Principle

This method uses a colourimetric approach for the quantification of reducing carbohydrates. 3,5-dinitrosalicylic acid or DNSA (yellow) is used to check for the presence of a free carbonyl group present exclusively in reducing sugars of the carbohydrate family. The reagent, when reduced, forms an orange-red compound called 3-Amino-5-nitrosalicylic acid or ANSA, which absorbs light strongly at 540nm.

An aromatic reagent called DNSA (yellow), when reacts with reducing sugars, forms ANSA (orange), which absorbs light at 540 nm. This happens by the reduction of the 3-nitro group of DNSA to an amino group see Fig. (**3.5**) [5]. Simultaneously, the aldehyde and ketone group (free carbonyl group) of the reducing sugar is oxidised. This method can therefore be used to determine the activity of several enzymes like amylase, invertase, *etc.*, that involve the production of reducing sugars.

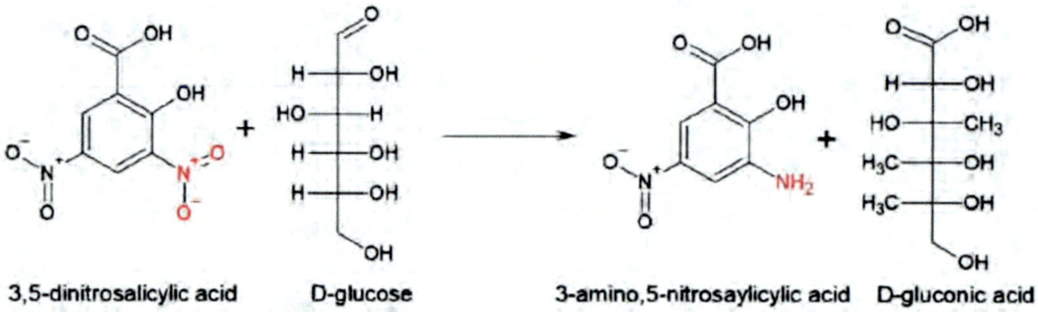

Fig. (3.5). DNSA reaction.

Range of detection: 0.2 - 2 mg/mL of glucose.

Requirements

• Equipment - Magnetic Stirrer

• Chemicals - DNSA Reagent, Sodium potassium tartrate solution (30 g of sodium potassium tartrate in 50mL distilled water), 2 N sodium hydroxide (2N NaOH), Dinitro salicylic acid (DNSA), Distilled water, Sample.

• Glucose stock solution - 2mg glucose per mL distilled water.

• Glassware- Test-tube, Test-tube stand, Pipettes, Beaker, Test tube caps, Tissue paper, Wash bottle.

Procedure

1. From a series of 8 test tubes: Consider the first tube as blank and label the rest as 1-7; tubes 1 to 6 will be used for the construction of a standard curve. Test tube 7 is for the unknown sample.

2. Pipette out into the test tubes 1 to 6 different volumes of stock solution (1.80mg /ml) and make up the volume up to 1mL with distilled water (Table **3.2**).

3. Add 3 ml of DNSA reagent to all eight test tubes and mix well.

4. Place test tubes 1-7 in a boiling water bath for 15 minutes.

5. After cooling to room temperature (use a cold-water bath if necessary), record the absorbance with a spectrophotometer at 540 nm.

6. Take the absorbance (OD) of Blank first to make it zero, followed by all the tubes 1-7. (Caution: Wash the cuvettes with distilled water each time after taking OD).

7. Prepare the standard curve and calculate the glucose concentration in the sample (Fig. **3.6**).

Table 3.2. Reference data for the DNSA method.

Sr. No.	Conc of Glucose (mg/ml)	Stock Volume (ml)	Distilled Water (ml)	DNSA (ml)		Absorbance (at 540 nm)
B	0	0	1	3	Incubate the Test Tubes in a Boiling Water Bath for 15 Mins	0
1	0.36	0.2	0.8	3		0.12
2	0.72	0.4	0.6	3		0.25
3	1.08	0.6	0.4	3		0.33
4	1.44	0.8	0.2	3		0.47
5	1.80	1	0	3		0.55
6	?	1ml (diluted 200 times)	-	3		0.28

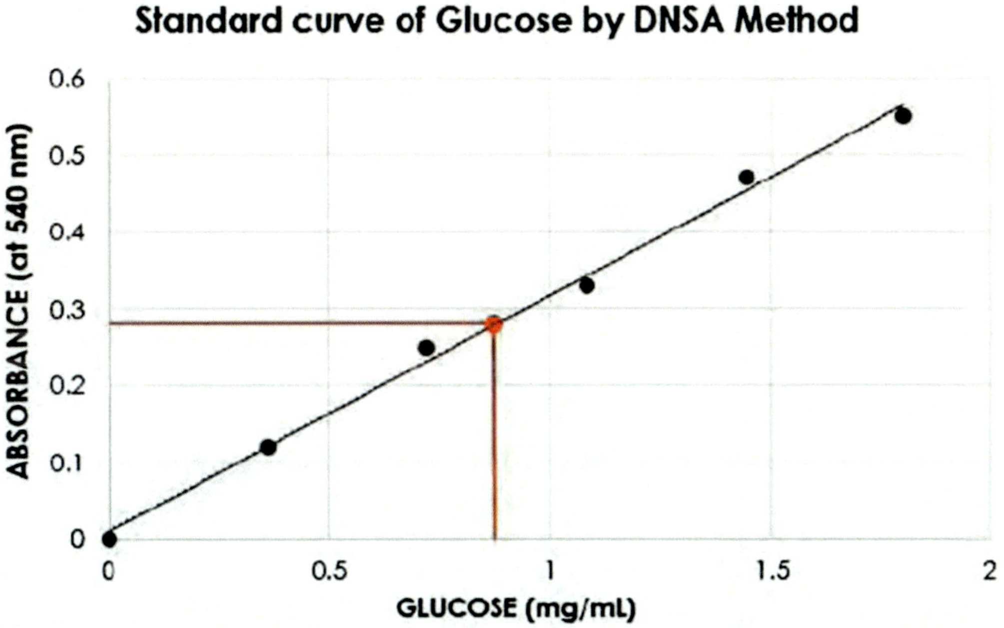

Fig. (3.6). Standard graph of glucose by DNSA Method. The standard graph provided is only for representational purpose. Data shown in this graph may differ.

Results

Example of a glucose standard curve for DNSA method of sugar estimation.

Calculation:

1. Follow the Eqn. 3 for the calculation of unknown sample concentration.

Example:

The equation thus obtained: $y = 0.3079x + 0.0095$

Concentration of unknown (diluted) (x) = 0.87 mg /ml

Concentration of unknown (dilution factor * x) = 20 * 0.87 = 174 mg /ml = 1740 mg/dL

Did you know?

- All monosaccharides and disaccharides (except sucrose) are reducing sugars.

FOLIN-WU METHOD

Principle

The Folin-Wu method was developed in 1953 for determining the concentration of sugar present in the blood. In this assay, the cupric ions present in the alkaline reagent are reduced to cuprous ions or cupric sulphate is converted to cuprous oxide by glucose present in the sample of blood [6]. This, in turn, reduces the phosphomolybdic acid to phosphomolybdous acid, giving blue colour intensity, which is measured at 420 nm. Usually, the Folin-Wu Method depends on the heating time, extent of temperature and degree of alkalinity (Fig. **3.7**) [6].

Requirements

Equipment- Bunsen Burner, Water Bath

Chemicals- 1% sugar solution, Alkaline copper solution, Phosphomolybdic acid, 10% sodium tungstate, 2/3N Sulphuric acid, 10% Phosphoric acid solution, 10% Sodium hydroxide solution

Glassware- Test tubes, Folin-Wu tubes

Others: Test tube racks, Distilled water (D/W), Whatman filter paper no. 1

$$Glucose + Cu^{2+} \longrightarrow CuOH \xrightarrow{\Delta} Cu_2O$$

$$Cu_2O + \text{Phosphomolybdic acid} \longrightarrow \text{Omlybdenum Blue}$$

$$Glucose + Cu^{2+} \longrightarrow CuOH \xrightarrow{\Delta} Cu_2O$$

$$Cu_2O + \text{Phosphomolybdic acid} \longrightarrow \text{Omlybdenum Blue}$$

Fig. (3.7). Folin-Wu reaction.

Procedure

1. Take 0.5 ml of blood to be tested in a boiling tube containing 3.5 ml of D/W.

2. Add 0.5ml of 10% sodium tungstate and 0.5ml of 2/3N H_2SO_4 while shaking the tube.

3. Allow it to rest for 10 min.

4. Filter the contents out using Whatman filter paper no. 1.

5. The filtrate obtained is called tungstic acid blood and is taken as the test sample.

6. Transfer 2 ml of this filtrate to the Folin-Wu tube graduated at 25ml mark. Label it as tube 1.

7. Transfer 2 ml of standard glucose solution and 2 ml water each as blank to the other two Folin-Wu tubes. Label them as tubes 2 and 3, respectively.

8. Add 2 ml of alkaline $CuSO_4$ to all 3 tubes.

9. Place these tubes in a boiling water bath for 8 min.

10. Cool the tube under running tap water.

11. Add 2 ml of phosphomolybdic acid to each tube. Incubate it for 1 min.

12. Dilute with D/W up to the graduated mark on the Folin-Wu tubes.

13. Note the absorbance at 420 nm.

Results

______mg of sugar is present in the tested blood sample.

Note: Normal value of blood sugar level – 70-120mg/dL of blood.

Calculations:

Note the following values:

mg of glucose in standard=

OD of test=

OD of standard=

Therefore,

$$mg\ of\ glucose/100ml\ of\ blood = mg\ og\ glucose\ in\ standard\ X\ \frac{OD\ of\ test}{OD\ of\ standard}\ X\ \frac{100}{0.2}$$

HEXOKINASE METHOD

Principle

The hexokinase (HK) method is an accurate method of sugar estimation present in the plasma. To determine the sugar present in plasma or serum, they are first deproteinized to obtain a clear supernatant, which is used for the reaction. Here, if glucose is present in the sample, it will react to ATP (Adenosine Triphosphate) with the help of the HK enzyme to form glucose-6-phosphate. This glucose- 6-phosphate then is catalysed by the enzyme glucose-6-phosphate dehydrogenase (G6PD), in the presence of NAD+ or NADP+ (nicotinamide adenine dinucleotide phosphate) to form NADH or NADPH (Dihydronicotinamide-adenine dinucleotide phosphate) and 6-phosphogluconate [7, 8]. The amount of NADH or NADPH formed is proportional to the concentration of glucose present in the sample, which is determined by measuring the amount of NADH or NADPH generated by measuring the absorbance at 340nm (Fig. **3.8**) [7, 8].

This method is considered highly accurate and precise, due to which, it is widely used for plasma glucose estimation.

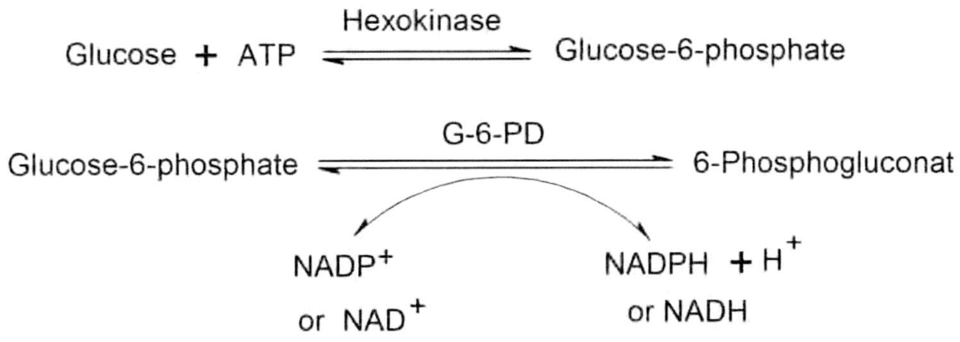

Fig. (3.8). Hexokinase method reaction.

Requirements

Equipment- Spectrophotometer, Micropipette

Chemicals- Standard glucose solution

Glassware- Cuvettes

Others: Distilled water (D/W)

Procedure

1. Label 3 cuvettes, "Blank, Standard and Sample".

2. Add 2 ml of HK enzyme to all 3 cuvettes.

3. Pipette 20 µl of D/W and add in Blank cuvette.

4. Pipette 20 µl of 10mmol/l glucose standard solution and add in Standard labelled cuvette.

5. Now add 20 µl of serum or plasma sample directly OR deproteinized plasma or serum sample into the cuvette labelled sample.

6. Gently mix the contents of the cuvette using a pipette.

7. Transfer all 3 cuvettes to a spectrophotometer and read the absorbance at 340 nm.

8. A graph plot of absorbance vs glucose concentration will be obtained, which will be used to determine the concentration of the sample.

Results

Example of a standard curve for Hexokinase Method of plasma sugar estimation (Fig. **3.9**).

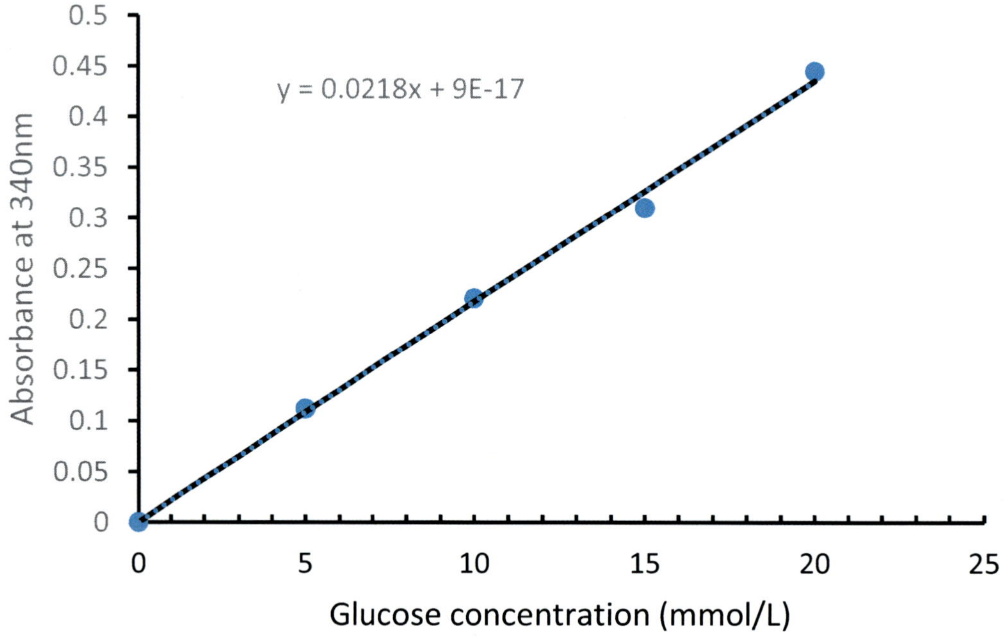

Fig. (3.9). Standard curve of the Hexokinase method. The standard graph provided is only for representational purpose. Data shown in this graph may differ.

Calculation:

Follow the Eqn. 3 for the calculation of unknown sample concentration.

PHENOL-SULFURIC ACID METHOD

Principle

The Phenol-Sulfuric Acid method is a colourimetric method for the determination of the total concentration of carbohydrates present in a given sample. This method detects all classes of carbohydrates, *i.e.*, monosaccharides, disaccharides, oligosaccharides and polysaccharides, but the absorptivity of each different class varies [9, 10]. Usually, the sample is prepared in such a way that it is a clear aqueous solution to which phenol and sulphuric acid are added. The concentrated sulphuric acid breaks down any complex classes (poly-, oligo- & disaccharides) of carbohydrates to monosaccharides, which are further dehydrated to furfural and hydroxymethyl furfural from pentose and hexose sugars, respectively [9, 10, 14]. These compounds then condense with phenol to form stable yellow-gold colour that can be measured spectrophotometrically at 490nm [9, 10].

This experiment requires one to create a standard glucose curve to determine the carbohydrate concentration of the sample (Fig. **3.10**).

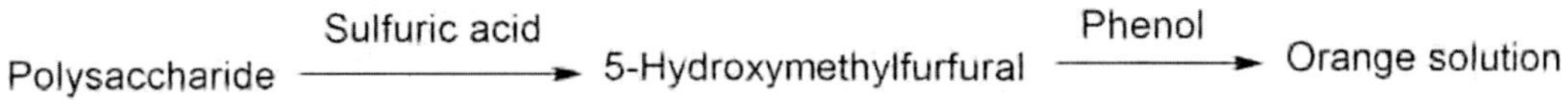

Fig. (3.10). Phenol-Sulfuric acid method reaction.

Requirements

Equipment- Spectrophotometer, Vortex mixer, Water bath (should be maintained at 25°C)

Chemicals- 80% (wt/wt) Phenol solution, Glucose standard solution 100mg/L, Concentrated Sulphuric acid.

Glassware- Test tube, Test-tube stand, Pipettes

Others: Distilled water (D/W)

Procedure

1. Prepare standard glucose solution for creating a standard curve, such that the tubes used to create a standard curve will have 0-100 µg glucose/2 ml.

2. Prepare the sample tube by diluting it such that the tested sample will contain 20-100µg glucose/ 2ml.

3. In a small test tube, pipette 1ml of sample and add 1ml of D/W.

4. Add 0.05 ml of 80% phenol to the tube and mix the sample well using a vortex mixer.

5. Add 5ml of concentrated sulphuric acid rapidly to the tube. Mix the sample well using a vortex mixer.

6. Let the tube stand for 10min and then place it in a water bath at 25°C for 10min.

7. Vortex the tubes again before reading absorbance.

8. Read the absorbance of the sample at 490nm.

Results

Example of a glucose standard curve for Phenol-Sulphuric Acid Method of sugar estimation (Fig. **3.11**).

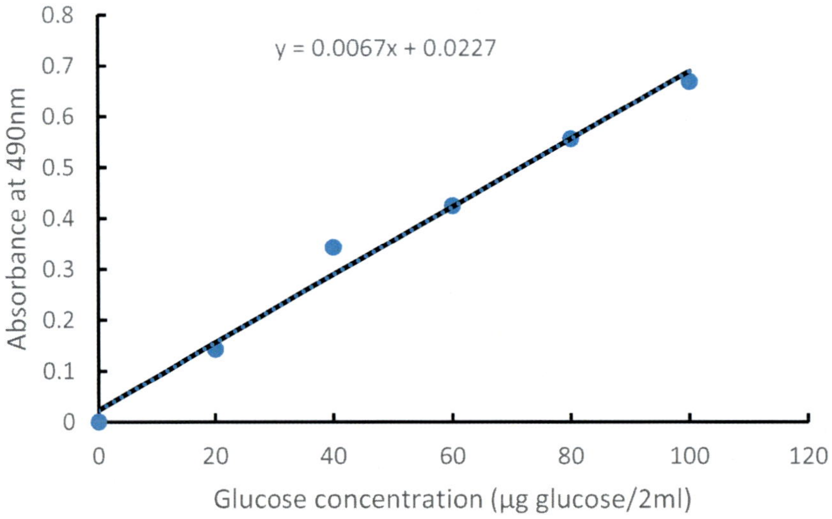

Fig. (3.11). Standard curve of Glucose by Phenol- Sulphuric Acid Method. The standard graph provided is only for representational purpose. Data shown in this graph may differ.

Calculation:

Follow the Eqn. 3 for the calculation of unknown sample concentration.

SOMOGVI-NELSON METHOD

Principle

To determine the amount of reducing sugars in a given sample the Somogyi-Nelson method is most often used. In this method, reducing sugars reduce Cu(II) ions to Cu(I) ions, which is the base of this method. The reaction usually takes place in an alkaline solution, which contains citrate ions and tartrate that help to keep Cu ions in the solution [11, 12].

The Cu(I) ions formed by the reducing sugars, further reduce arsenomolybdate complex present in the Nelson reagent (which is prepared by reacting sodium arsenate (Na_2HAsO_7) and ammonium molybdate [$(NH_4)_6Mo_7O_{24}$] in sulphuric acid (H_2SO_4)) which when is added to the sample to tested. The reduction of this arsenomolybdate complex produces a stable blue colour due to molybdenum blue formation, which is measured spectrophotometrically at 520nm [11]. The colour intensity formed tells us about the amount of sugar present in the sample, which can be determined through a standard curve of sugars (Fig. **3.12**) [12].

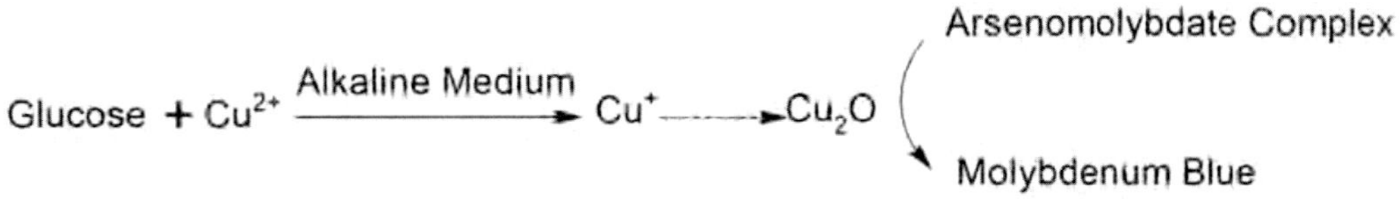

Fig. (3.12). Somogvi-Nelson method reaction.

Requirements

Equipment- Spectrophotometer, Vortex mixer, Water bath at 100°C.

Chemicals- Copper reagent, Nelson's arsenomolybdate reagent, Glucose standard solutions (0, 0.02, 0.04, 0.06, 0.08,0.1 (mg/ml))

Glassware- Test-tube, Test-tube stand, Pipettes, cuvettes

Others: Distilled water (D/W)

Procedure

1. Prepare glucose standards with D/W.

2. Add 1 ml of a standard solution prepared in different tubes.

3. Add 1ml D/W to the tubes used as blank.

4. Prepare the sample to be tested appropriately by diluting it.

5. To each tube, add 1 ml of Cu reagent and mix.

6. Heat the solutions over a water bath of 100°C for 10-20min.

7. Let the tubes cool to room temperature.

8. Add 1 ml of Nelson's reagent to the tubes.

9. Measure the absorbance at 520nm.

Results

Example of a glucose standard curve for Somogyi-Nelson Method of sugar estimation (Fig. **3.13**).

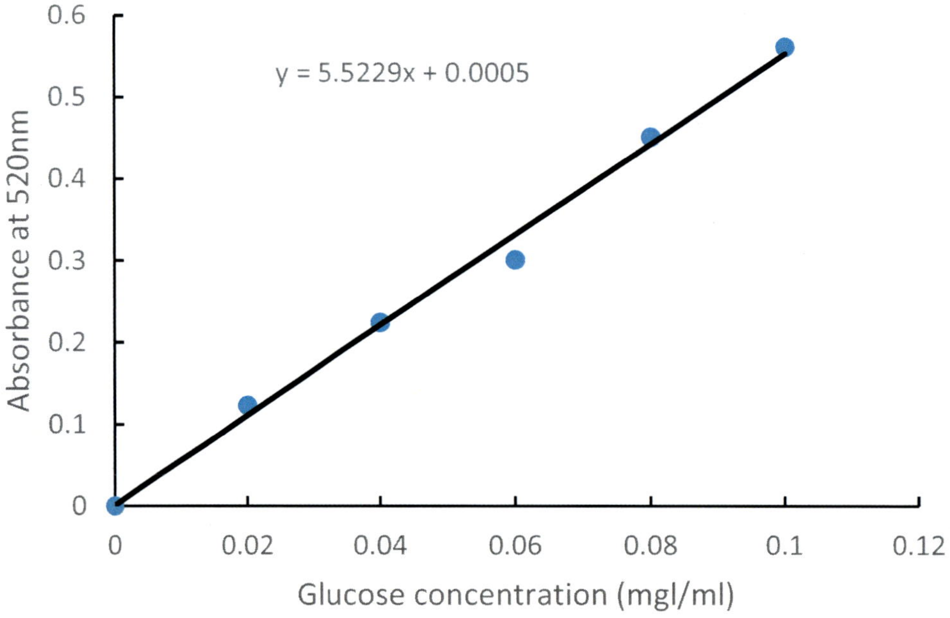

Fig. (3.13). Standard curve for Somogvi-Nelson method. The standard graph provided is only for representational purpose. Data shown in this graph may differ.

Calculation:

Follow the Eqn. 3 for the calculation of unknown sample concentration.

GOPOD METHOD

Principle

GOPOD/GODPOD or Glucose oxidase/peroxidase/dye method is an enzymatic method of determination of the concentration of D-glucose which uses two enzymes. The first enzyme, glucose oxidase, oxidizes D-glucose to D-glucono-1,5-lactone and releases hydrogen peroxide as a by-product. The amount of D-glucose present is measured through the reaction of hydrogen peroxide with peroxidase and a leuco dye- a colourless compound (*e.g.*, 4-aminoantipyrine and phenol) that gets converted to a coloured (red-coloured quinoneimine dye) compound when oxidized [12 - 14]. The peroxidase acts as a catalyst in the reaction to convert the leuco dye to a coloured compound which is then measured spectrophotometrically. Various types of dyes are available commercially nowadays that act as leuco- dyes (Fig. **3.14**) [12 - 14].

$$\text{Glucose} + H_2O + O_2 \xrightarrow{\text{Glucose Oxidase (GOD)}} \text{Gluconic Acid} + H_2O_2$$

$$H_2O_2 + \text{Phenol} + \text{4-aminoantipyrine} \xrightarrow{\text{Peroxidase (POD)}} \text{Red quinoneimine dye} + H_2O$$

Fig. (3.14). GOPOD method reaction.

Requirements

Equipment- Colourimeter, Water Bath, Micropipettes

Chemicals- Glucose standard solution (100mg/dl), GOD-POD reagent

Glassware- Test tubes

Others- Disposable tips, Test tube racks, D/W

Procedure

1. Take 3 test tubes and label them as Blank(B), Test(T) and Standard(S).

2. In the B test tube, add 10 µl of D/W and 1 ml of GOD-POD reagent.

3. In the S tube, add 10 µl of glucose standard and 1 ml of GOD-POD reagent.

4. In the T tube, 10 µl of sample and 1 ml of GOD-POD reagent.

5. Mix all the 3 tubes well and incubate them at 37°C for 10 minutes in a water bath.

6. Measure the absorbance of the standard and test sample at 540 nm (if 4-aminoantipyrine and phenol are used as leuco dye) against blank within 60 minutes.

Results

The Concentration of D-glucose in the unknown sample is ______ (mg/dl).

Calculator:

$$Concentration\ of\ Glucose\ (mg/dl) = \frac{Absorbance\ of\ Test}{Absorbance\ of\ Sample}\ X\ 100$$

CONCLUSION

The quantitative determination of carbohydrate assays that are discussed in this chapter is based on colourimetric methods. Depending upon the type of analysis for carbohydrates to be quantified, an appropriate method needs to be selected.

VIVA VOCE

1. What is the anthrone reagent method?

Ans. The reagent is anthranol, the enol tautomer of anthrone, which reacts by condensing with the carbohydrate furfural derivative to give a green colour in dilute and a blue colour in concentrated solutions, which is determined colourimetrically. The blue-green solution shows an absorption maximum of 620 nm.

2. What is the principle behind the Molisch and anthrone test?

Ans. Principle of Molisch's Test: To detect the presence of carbohydrates, the solution is first treated with a strong acid. This is for hydrolyzing the carbohydrate to monosaccharide. A compound named furfural is then made when water is removed from monosaccharides.

3. What does anthrone reagent contain?

Ans. Test solution: 5% Glucose, 5% Sucrose, 5% Starch.

4. Who discovered the anthrone method?

Ans. The detection of carbohydrate material using anthrone (9-Keto anthra- cene) was introduced by Dreywood in 1946.

5. Can sucrose be estimated with the DNSA method?

Ans. Sucrose is a disaccharide that has to be first hydrolyzed using either enzyme like invertase or undergoing acidic hydrolysis to give glucose and fructose monosaccharides. The DNSA method can now be used to determine the total reducing sugars.

6. What is a DNSA reagent?

Ans. 3,5-Dinitrosalicylic acid (DNS or DNSA, IUPAC name 2-hydroxy-3-5-dinitrobenzoic acid) is an aromatic compound that reacts with reducing sugars and other reducing molecules to form 3-amino-5-nitrosalicylic acid, which strongly absorbs light at 540 nm.

7. Why is the DNSA method used?

Ans. DNSA is more sensitive and easier to use than Benedict's reagent. This method also helps to measure the activity of enzymes such as invertase, cellulase and amylase where reducing sugars are produced.

8. What are the absorption maxima for the estimation of reducing sugar by the DNSA method?

Ans. DNSA reacts with a free carbonyl group of the reducing sugar under alkaline conditions, forming 3-amino-5-nitrosalicylic acid, with maximum absorption at 540 nm, allowing a quantitative spectrophotometer measurement of the amount of reducing sugar present.

9. On what parameters does the Folin-Wu method depend?

Ans. The Folin-Wu Method depends on the heating time, extent of temperature and degree of alkalinity.

10. In the Folin-Wu method, what component of alkaline reagent is reduced?

Ans. Cupric ions in alkaline reagents are reduced to cupric ions.

11. What method is used for sugar estimation in plasma?

Ans. The Hexokinase method is used to measure sugar in plasma.

12. At what wavelength is absorbance measured in the Hexokinase method?

Ans. At 340nm, absorbance is measured, which determines the presence of NADPH or NADH, which is proportional to the amount of sugar present in plasma.

13. In Phenol-Sulfuric Acid method, what breaks down complex classes such as poly-, oligo- and disaccharides?

Ans. The concentrated sulphuric acid breaks down complex classes (poly-, oligo- & disaccharides) of carbohydrates to monosaccharides.

14. How is molybdenum blue formed in the Somogyi-Nelson method?

Ans. The reduction of arsenomolybdate complex due to Cu(I) ions forms a molybdenum-blue complex.

15. What method is used to determine the presence of D-glucose in the sample?

Ans. GOPOD/GODPOD or Glucose oxidase/peroxidase/dye method, an enzymatic method, is used for the determination of the concentration of D-glucose.

16. What are the two enzymes used in the GOPOD/GODPOD method?

Ans. Glucose oxidase and peroxidase are the two enzymes used in the GODPOD method.

REFERENCES

[1] Beck, A.; Hunt, K.; Carlson, R. Measuring cellular biomass composition for computational biology applications. *Processes,* **2018**, *6*(5), 38. [http://dx.doi.org/10.3390/pr6050038]

[2] Bhaumik, P.; Dhepe, PL. Recent advances and future projections on sustainable method development for furfural production. *Bull. Catal. Soc. India.,* **2016**, *14*, 1-32.

[3] Zhou, C.; Zhao, J.; Yagoub, AE.; Ma, H.; Yu, X.; Hu, J.; Bao, X.; Liu, S. Conversion of glucose into 5-hydroxymethylfurfural in different solvents and catalysts: Reaction kinetics and mechanism. *Egypt. J. Pet.,* **2017**, *26*(2), 477-487.

[4] Haldar, D.; Sen, D.; Gayen, K. Development of spectrophotometric method for the analysis of multi-component carbohydrate mixture of different moieties. *Appl. Biochem. Biotechnol.,* **2017**, *181*(4), 1416-1434. [http://dx.doi.org/10.1007/s12010-016-2293-3] [PMID: 27787768]

[5] McKee, LS Measuring enzyme kinetics of glycoside hydrolases using the 3, 5-dinitrosalicylic acid assay. *Protein-carbohydrate interactions: Methods and protocols,* **2017**, 27-36.

[6] Jain, A.; Jain, R.; Jain, S. Basic Techniques in Biochemistry. In: *Microbiol. Mol. Biol*; Springer: New York, NY, USA, **2020**.

[7] Duxbury, M. An enzymatic clinical chemistry laboratory experiment incorporating an introduction to mathematical method comparison techniques. *Biochem. Mol. Biol. Educ.,* **2004**, *32*(4), 246-249. [http://dx.doi.org/10.1002/bmb.2004.494032040366] [PMID: 21706732]

[8] Sonagra, AD; Motiani, A *Hexokinase Method*; StatPearls Publishing LLC.: Treasure Island, **23**.

[9] Nielsen, SS Phenol-sulfuric acid method for total carbohydrates. *Food analysis laboratory manual,* **2010**, 47-53. [http://dx.doi.org/10.1007/978-1-4419-1463-7_6]

[10] Nielsen, SS; Nielsen, SS Total carbohydrate by phenol-sulfuric acid method. *Food analysis laboratory manual,* **2017**, 137-141. [http://dx.doi.org/10.1007/978-3-319-44127-6_14]

[11] Romadhoni, R.P.; Purbaningtias, T.E.; Muhaimin, F.A. Determination of reduction sugar form banana (musa acuminata musa balbisiana colla) with different cooking process by uv-visible

spectrophotometer. *Conference: The 2nd International Seminar on Chemical Education,* September 2017, Yogyakarta, Indonesia, pp. 403-409.

[12] Burrin, J.M.; Price, C.P. Measurement of blood glucose. *Ann. Clin. Biochem.,* **1985**, *22*(4), 327-342. [http://dx.doi.org/10.1177/000456328502200401] [PMID: 3898972]

[13] Dandekar, S.P.; Rane, S.A. *Practicals and Viva in Medical Biochemistry*; Elsevier: New Delhi, **2004**.

[14] BeMiller, J.N. Carbohydrate analysis. *Food analysis,* **2017**, 333-60.

CHAPTER 4

Introduction to Proteins

Sinchan Hait[1,*] and **Sai Joshi**[2]

[1] *Department of Chemical Engineering, BITS-Pilani K.K. Birla Goa Campus, Goa-403726, India*

[2] *CSIR-NCL, Dr. Homi Bhabha Road, Pune - 411 008, India*

Abstract: Proteins are the ambidexterity macromolecules which play a crucial role in biological processes. Through proteins, genetic information is expressed. They exhibit enormous functions which are crucial in various pathways. Some proteins act as an enzyme which is vital for catalyzing various metabolic reactions in living cells. All proteins are made up of amino acids which join together via a peptide bond to form a polypeptide chain which is a precursor for the further development of protein. A nascent protein undergoes several post-translational modifications to achieve its true function. Based on the amino acids present in the protein, its function determines its role as hormones, enzymes, antibodies, transporters, *etc*. In this chapter, a brief introduction to protein's structure and function is discussed.

Keywords: Amino acids, Antibodies, Enzyme, Gene expression, Hormone, Peptide bond, Protein, Polypeptide chain, Post-translation modification, Transporters.

INTRODUCTION

Proteins are one of the most important macromolecules, which are abundant in living organisms. The genetic information in nucleic acids is expressed *via* proteins. However, these macromolecules are constructed from 20 amino acids. In addition, depending upon these amino acid combinations, proteins can be enzymes, hormones, antibodies, transporters, muscle fibre, *etc*. In humans, the Recommended Dietary Allowance (RDA) for protein is 0.8 grams of protein per kilogram of body weight. Some of the major sources of proteins are milk, fish, meat, beans, *etc*. Proteins are composed of amino acids. A basic structure of amino acids contains an amino group (-NH_2) and a carboxylic acid group (-COOH) (Fig. **4.1**) [1, 2].

* **Corresponding author Sinchan Hait:** Department of Chemical Engineering, BITS-Pilani K.K. Birla Goa Campus, Goa-403726, India; E-mail: sinchan.hait1@gmail.com

Pamela Jha (Ed.)

These amino acids are joined by peptide bonds to form proteins (Fig. **4.2**). There are 20 amino acids present, these amino acids are mainly present in zwitterions (Fig. **4.3**) form, and they (except glycine) exhibit optical activity due to asymmetric α-carbon atoms (Fig. **4.4**).

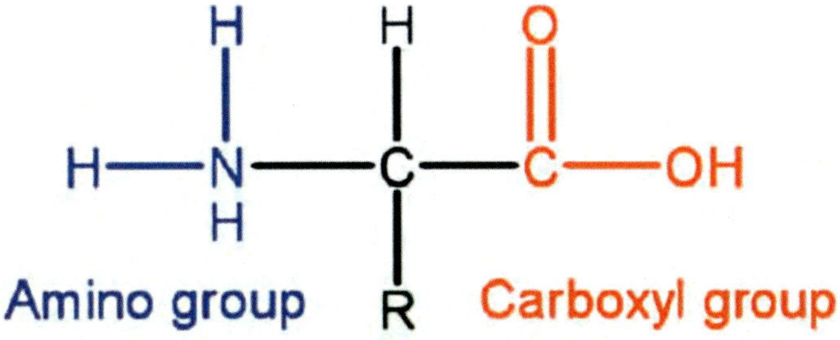

Fig. (4.1). Amino acid structure.

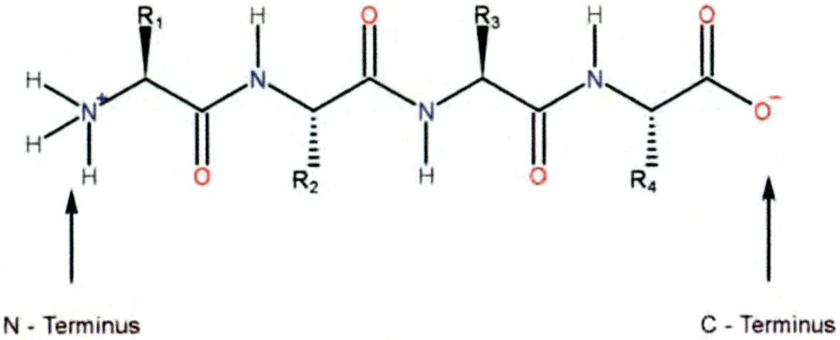

Fig. (4.2). Peptide chain.

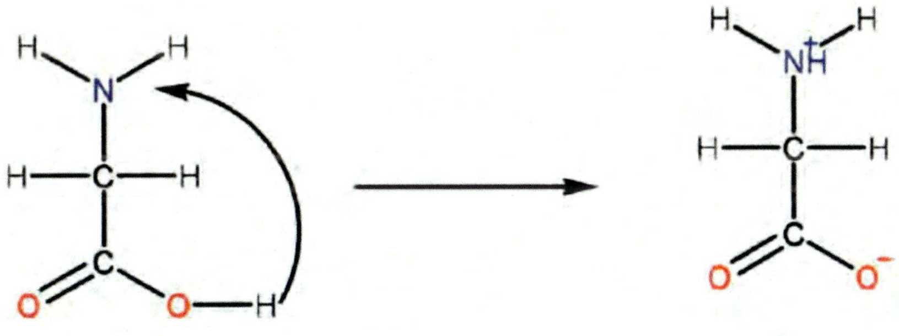

Fig. (4.3). Zwitterion.

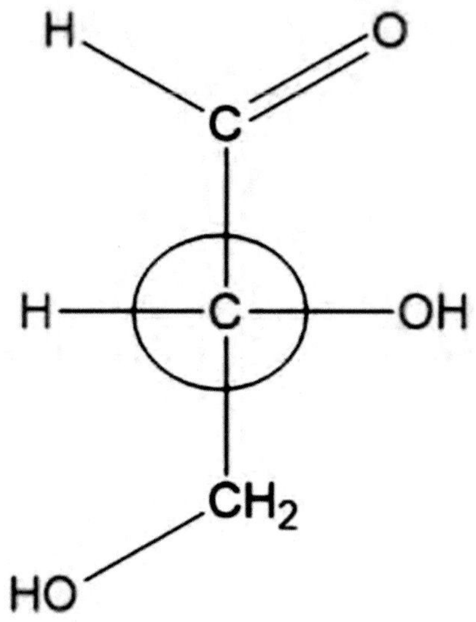

Fig. (4.4). Chiral carbon.

There are several methods available to determine amino acids, which help us to identify the protein.

PROTEIN STRUCTURE

The amino acid sequence of a protein determines its structural configuration and its function. A protein may contain 20 different amino acids (Table **4.1**) for Amino acid properties and convention, each containing an amine group and an acid group at the 2 terminals, with a distinct side chain.

Table 4.1. Amino acid properties and convention.

Amino Acid	Abbreviation	M_r	pKa Values			pI	Hydropathy Index	Occurrence in Protein (%)
			pK_1 (-COOH)	pK_2 (-NH_3^+)	pK_R (R group)			
Nonpolar, Aliphatic								
Glycine	Gly, G	75	2.34	9.60	-	5.97	-0.4	7.2
Alanine	Ala, A	89	2.34	9.69	-	6.01	1.8	7.8
Proline	Pro, P	115	1.99	10.96	-	6.48	1.6	5.2
Valine	Val, V	117	2.32	9.62	-	5.97	4.2	6.6
Leucine	Leu, L	131	2.36	9.60	-	5.98	3.8	9.1
Isoleucine	Ile, I	131	2.36	9.68	-	6.02	4.5	5.3

(Table 4.1) cont.....

Amino Acid	Abbreviation	M_r	pKa Values			pI	Hydropathy Index	Occurrence in Protein (%)
			pK_1 (-COOH)	pK_2 ($-NH_3^+$)	pK_R (R group)			
Methionine	Met, M	149	2.28	9.21	-	5.74	1.9	2.3
Aromatic								
Phenylalanine	Phe, F	165	1.83	9.13	-	5.48	2.8	3.9
Tyrosine	Tyr, Y	181	2.20	9.11	10.7	5.66	-1.3	3.2
Tryptophan	Trp, W	204	2.38	9.39	-	5.89	-0.9	1.4
Polar, Uncharged								
Serine	Ser, S	105	2.21	9.15	-	5.68	-0.8	6.8
Threonine	The, T	119	2.11	9.62	-	5.87	-0.7	5.9
Cysteine	Cys, C	121	1.96	10.28	8.18	5.07	2.5	1.9
Asparagine	Asn, N	132	2.02	8.80	-	5.41	-3.5	4.3
Glutamine	Gln, Q	146	2.17	9.13	-	5.65	-3.5	4.2
Positively Charged								
Lysine	Lys, K	146	2.18	8.95	10.53	9.74	-3.9	5.9
Histidine	His, H	155	1.82	9.17	6.00	7.59	-3.2	2.3
Arginine	Arg, R	174	2.17	9.04	12.48	10.76	-4.5	5.1
Negatively Charged								
Aspartate	Asp, D	133	1.88	9.60	3.65	2.77	-3.5	5.3
Glutamate	Glu, E	147	2.19	9.67	4.25	3.22	-3.5	6.3

Based on the structure, proteins can be classified as [1]:

1. Primary structure- It contains a long amino acid sequence along the peptide chain, which is joined by peptide bonds (Fig. **4.5**).

2. Secondary structure- It refers to the local conformation of some part of the polypeptide. The formation of a secondary structure occurs due to the repeating pattern of hydrogen bonds between backbone atoms (Fig. **4.5**). Secondary structure includes:

• α-helix, the simplest arrangement of the polypeptide chain with its rigid peptide bonds.

• β-sheet, the backbone of the polypeptide is in a zig-zag pattern.

3. Tertiary structure- The 3-D arrangement of all amino acids in proteins is referred to as the tertiary structure. The location, direction and angle of bends are determined by bend-producing amino acid residues (Pro, Thr, Ser and Gly) (Fig. **4.5**).

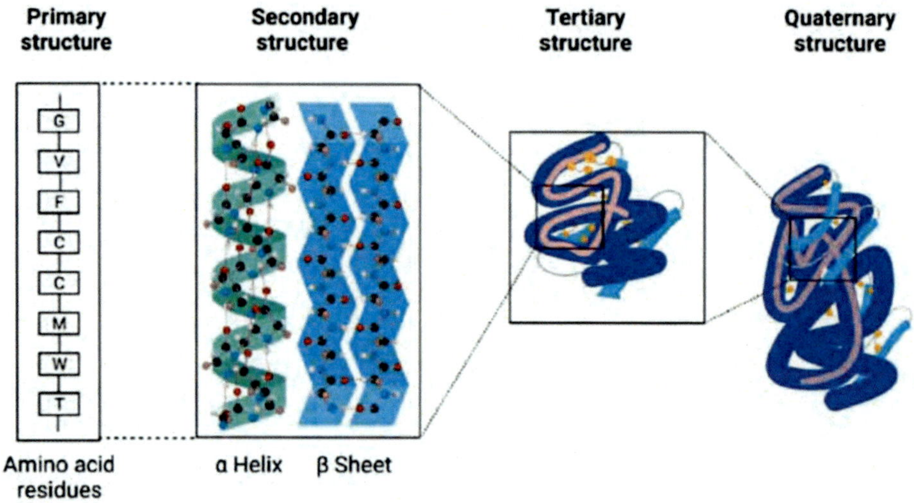

Fig. (4.5). Protein structure.

4. Quaternary structure- Some proteins contain two or more polypeptide chains, which may or may not be identical to each other; such arrangements in 3-D complexes constitute the quaternary structure (Fig. **4.5**).

The higher levels of structure of proteins are classified as:

1. Fibrous protein-The polypeptide chains are arranged in long stands or sheets. They provide structural support, shape and external protection to vertebrates.

Ex.- Keratin, Collagen.

2. Globular protein - The polypeptide chains are folded in a globular or spherical shape and contain several secondary structures. The enzymes and regulatory proteins are globular.

Ex.- Albumin, Histones.

FUNCTIONS OF PROTEIN

Proteins play many important roles in our body, a few of them are enumerated as follows [1, 2]:

Enzymes

All the enzymes present in our body are proteinaceous, however, ribozymes are an exception.

Membranes and Transport Molecules

Proteins form the major constituents of the bio-membrane, some of which also function as carriers or translocators, thus facilitating selective transport of the metabolites across the cell membranes.

Antibodies and Toxins

Antibodies, some toxins, and antibiotics are either proteins or peptides and form an integral part of the defense systems of living organisms.

Functional Proteins

Proteins play many roles in our bodies. They majorly form storage proteins like zein in maize, and casein in milk. They also form hormones like insulin.

Structural Proteins

Proteins also form major structures like hair, nail, proteins of the cell wall and cell membranes.

A few amino acid properties are mentioned in Table **4.1**.

To analyze the protein and its type quantitatively and qualitatively (Table **4.2**), various tests are performed, which are explained in Chapters **6 and 7**. The protein purification techniques are also discussed.

Table 4.2. Different qualitative tests for amino acids.

Test	Reactants	Specificity of Test	Colour Obtained
Xanthoproteic test	Conc. HNO_3 in an alkaline solution	Tyrosine, Tryptophan, Phenylalanine	Dark yellow or orange
Millon's test	$Hg(NO_3)_2$ in HNO_3 and $NaNO_2$	Tyrosine	Red
Pauly's Diazo test	Diazotized sulfanilic acid in an alkaline solution	Tyrosine and Histidine	Red
Hopkin's-Cole test	Glyoxylic acid in Conc. H_2SO_4	Tryptophan	Purple colour ring

(Table 4.2) cont.....

Test	Reactants	Specificity of Test	Colour Obtained
Ehrlich's test	p-dimethylaminobenzaldehyde (DMAB)	Tryptophan	Blue-violet
Sakaguchi test	α- naphthol (1-hydroxy naphthalene) and hypobromite or hypochlorite in an alkaline solution	Arginine	Red
Lead sulfide test	Lead acetate in an alkaline solution	Cysteine	Black precipitate
Nitroprusside test	Sodium nitroprusside in an alkaline solution	Cystine	Red
Ninhydrin test	Ninhydrin	All amino acids	Purple (yellow colour for proline)

CONCLUSION

Proteins perform various functions in living organisms, from regulating to catalysing. Based on the amino acids present in the protein, its function and activity are determined, which makes it essential for determining the proteins qualitatively and quantitively.

VIVA VOCE

1. Give structural classification of proteins?

Ans. Based on the structure, proteins are classified as primary, secondary, tertiary, and quaternary.

2. What is the difference between fibrous protein and globular protein?

Ans. In fibrous proteins, polypeptide chains are arranged in long stands or sheets and provide structural support, *e.g.*, keratin.

Globular proteins are polypeptide chains folded in a globular or spherical shape and contain several secondary structures, *e.g.*, albumin.

3. What is the RDA of protein?

Ans. RDA of protein is 0.8 gm of protein per kilogram of body weight.

REFERENCES

[1] Nelson, D.L.; Lehninger, A.L.; Cox, M.M. *Lehninger principles of biochemistry*; Macmillan, **2008**.

[2] Walker, J.M., Ed. *The protein protocols handbook*; Springer Science & Business Media, **1996**. [http://dx.doi.org/10.1007/978-1-60327-259-9]

CHAPTER 5

Qualitative Tests for Protein Detection

Sinchan Hait[1,*]

[1] *Department of Chemical Engineering, BITS-Pilani K.K. Birla Goa Campus, Goa-403726, India*

Abstract: Qualitative tests are the primary assays for detecting the presence of any protein in a given sample. Qualitative detection of proteins is based on the chemical interactions between amino acids and certain chemical reagents added to them. These chemical reactions yield colour formation, which corresponds to specific amino acid or a group of amino acids present in the sample. In this chapter, some of such most important assays for qualitative analysis are discussed in detail.

Keywords: Amino acid, Assay, Protein.

INTRODUCTION

Qualitative tests for protein are based on the detection of amino acids present in it. Each amino acid has unique chemical properties, and some have common properties too. Under various chemical conditions, these amino acids react with certain chemical agents and develop colour compounds in them. This formation of colour compounds can then be correlated with amino acid/s present in a given sample. In a few cases, the colour formation can be due to some contamination in the sample, so further analysis is required, such as qualitative tests, SDS-PAGE, western blotting, *etc*., which are explained in the next chapters.

AROMATIC AMINO ACIDS (TYROSINE, TRYPTOPHAN)

Xanthoproteic Test

Aim

To determine the presence of proteins in a given sample by Xanthoproteic test.

* **Corresponding author Sinchan Hait:** Department of Chemical Engineering, BITS-Pilani K.K. Birla Goa Campus, Goa-403726, India; E-mail: sinchan.hait1@gmail.com

Principle

The aromatic amino acids, such as phenylalanine, tyrosine and tryptophan, contain phenolic groups. Under alkaline aromatic amino acids, concentrated

HNO_3 nitrates the phenyl ring and produces dark yellow or orange colour compounds (nitro-derivatives) (Fig. **5.1**) [1, 3]. But, the benzene ring in phenylalanine is not activated (in the case of tyrosine and tryptophan the benzene ring is active), and so it is not nitrated by nitric acid.

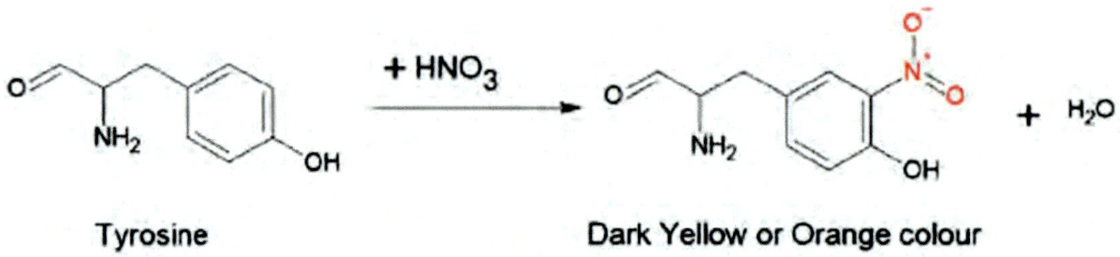

Fig. (5.1). Xanthoproteic test reaction.

Requirement

❖ ***Chemical***

- Conc. nitric acid
- 40% NaOH

❖ ***Glassware***

- Test tube
- 1 ml pipette
- Glass dropper

❖ ***Others***

- Test tube holder
- Bunsen burner
- Vortex

Procedure

1. Take 1 ml of the given sample in a test tube followed by a few drops of conc. HNO_3 and mix the contents.

2. Now boil the solution mix over a Bunsen burner for 2-3 minutes.

3. Allow the test tubes to cool down (Tip: if required use tap water to cool the content).

4. Add 0.1-0.3 ml of 40% NaOH solution and observe for colour change.

Results

❖ ***Positive Test***

The formation of yellow or orange colour means tyrosine and tryptophan are present.

❖ ***Negative Test***

No colour formation, which means the absence of tyrosine and tryptophan.

Precaution

1. Concentrated nitric acid is toxic and can cause serious injury to the skin. Prevent contact with the eye, clothing, and skin.

Application

• Xanthoproteic tests can be employed to identify aromatic amino acids (except phenylalanine).

Limitations

• Phenylalanine has a stable phenyl group; it does not react with HNO_3.

TYROSINE

Millon's Test

Aim

To determine the presence of protein in a given sample by Millon's test.

Principle

Millon's test is tyrosine specific. Hydroxybenzene radicals of tyrosine react with Millon's reagent in the presence of sodium nitrate ($NaNO_2$) to form a red-coloured complex (Fig. **5.2**) [1, 3]. Millon's reagent has mercuric sulphate and sulphuric acid.

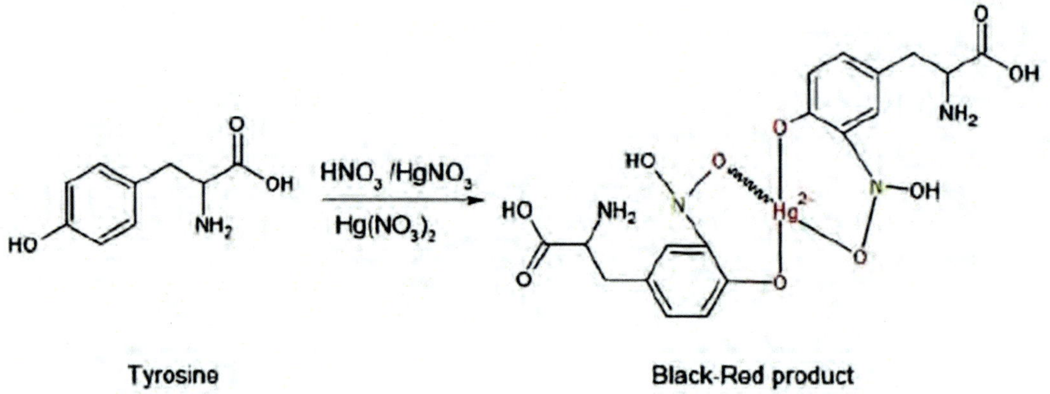

Fig. (5.2). Millon's test reaction.

Requirement

❖ *Chemical*

- Millon's reagent (Dissolve 15% of mercuric sulphate in 6N sulphuric acid)
- Sodium nitrate (1%)

❖ *Glassware*

- Test tube
- 1 ml pipette
- Glass dropper

❖ *Others*

- Test tube holder
- Boiling water bath
- Vortex

Procedure

1. Take 2 ml of a given sample in a test tube; to that, add 3 drops of Millon's reagent **(Caution: Millon's reagent is highly toxic and corrosive)** and mix gently.

2. Transfer the test tubes to a boiling water bath and incubate for 3-4 mins.

3. Cool the content and add a few drops of $NaNO_2$ solution.

4. Observe for any colour changes.

Results

❖ ***Positive Test***

The formation of red or pink colour precipitate indicates the presence of tyrosine.

❖ ***Negative Test***

No red or pink colour precipitate formation, which means the absence of tyrosine.

Precaution

Millon's reagent contains mercuric sulphate in sulphuric acid, and it is very toxic. The reagent must be kept away from fire; it may catch fire. Avoid inhaling and ingesting the reagent.

Application

• Helps to identify tyrosine-containing protein.

• This test is useful for casein protein and protein found in raw meat.

Limitations

• The presence of chlorine in the solution might interfere with the reaction.

• This test is exclusively for phenolic proteins, so the presence of any other phenolic compound in the sample and reaction mixer must be avoided.

TYROSINE AND HISTIDINE

Pauly's Diazo Test

Aim

To determine the presence of protein in a given sample by Pauly's Diazo test.

Principle

Only Tyrosine or Histidine responds to Pauly's Diazo test. The sulphanilic acid reagent contains sulphanilic acid in hydrochloric acid. Sulphanilic acid reacts with sodium nitrite and hydrochloric acid to form diazonium salt (Fig. **5.3**) [1]. This salt couples with either tyrosine or histidine and produces azo dye, a red-coloured chromogen.

$$\text{Sulphanilic acid} + HNO_2 + HCl \longrightarrow \text{Diazonium salt} + 2H_2O$$

$$\text{Diazonium salt} + \text{Tyrosine} \longrightarrow \text{Azo Dye} + HCl$$

Fig. (5.3). Pauly's Diazo test reaction.

Requirement

❖ ***Chemical***

- Sulphanilic acid reagent (1% of Sulphanilic acid prepared in 1N HCl)
- Sodium nitrite (5% W/V)
- Sodium carbonate (10%W/V)

❖ ***Glassware***

- Test tube
- 1ml pipette
- Glass dropper

❖ ***Others***

- Test tube holder
- Ice bucket

• Vortex

Procedure

1. Take 1 ml of sulphanilic acid reagent in a test tube and keep the test tube in a chilled environment (Ex.-ice bucket).

2. Add 0.1-0.3 ml of prechilled sodium nitrite solution and mix. Add a few drops of the pre-chilled amino acid solution immediately after and mix.

3. Then add sodium carbonate solution dropwise until the colour appears.

Results

❖ ***Positive Test***

The formation of red colour means tyrosine and histidine are present.

❖ ***Negative Test***

No colour formation, which means the absence of tyrosine and histidine.

Precaution

• Sodium nitrate is a strong oxidizer and can cause skin, eye, and respiratory tract irritation. Ingestion and inhalation must be avoided.

• Sulphanilic acid may cause severe skin irritation. Inhaling and ingesting the compound must be avoided.

Application

• Used to detect tyrosine and histidine-containing proteins, also allows for differentiation between the proteins containing histidine and tyrosine.

Limitation

• The formation of diazonium salts occurs at low temperatures thus it must be performed in the presence of ice.

• This test does not allow for differentiation between tyrosine and histidine, thus Millon's test must be performed.

TRYPTOPHAN

Hopkin's-Cole Test

Aim

To determine the presence of protein in a given sample by the Hopkin's-Cole test.

Principle

The Hopkins-Cole test is to detect tryptophan. In the presence of concentrated sulphuric acid, the indole moiety of tryptophan reacts with glyoxylic acid and gives rise to a purple-coloured product (Fig. **5.4**) [1].

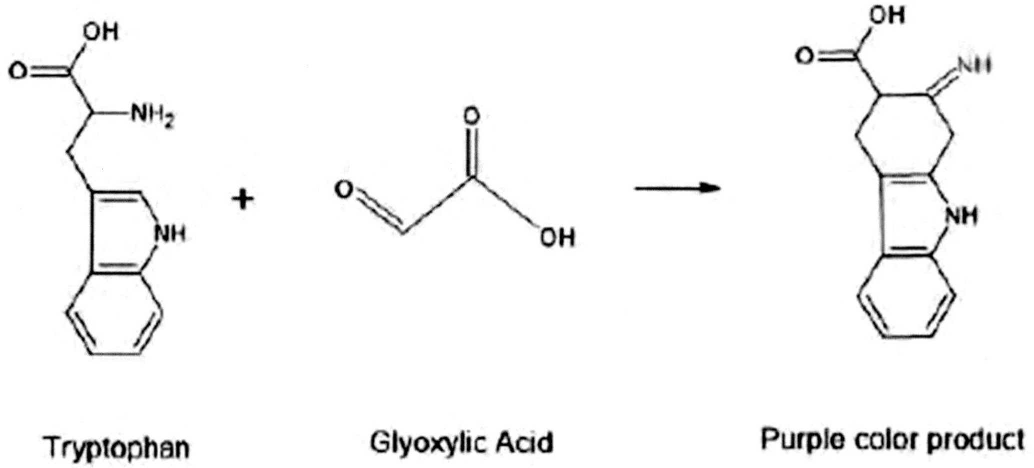

Fig. (5.4). Hopkin's-Cole test reaction.

Requirement

❖ ***Chemical***

• Acetic acid – Glyoxylic acid reagent (Glacial acetic acid is exposed to sunlight (for 5 – 6 hours) for the formation of small amounts of glyoxylic acid).

• Conc. sulphuric acid

❖ ***Glassware***

• Test tube

• 1 ml pipette

- Glass dropper

❖ ***Others***

- Test tube holder
- Boiling water bath
- Vortex

Procedure

1. Take 1 ml of the given sample in a test tube and mix it with 1 ml of acetic acid – glyoxylic acid reagent and mix it gently.

2. Hold the test tube at ~45°-50° and add conc H_2SO_4.

3. Observe for any colour changes.

Results

❖ ***Positive Test***

The formation of purple colour ring means tryptophan is present.

❖ ***Negative Test***

No purple colour ring formation, which means the absence of tryptophan.

Precautions

1. Hopkin's-Cole test reactions that need concentrated sulfuric acid are toxic and can cause severe burns. Avoid ingesting the substance.

2. Maintain caution for conc. H_2SO_4.

Application

- Performed to identify tryptophan.

Limitation

- The presence of chloride, chlorates and nitrates can interfere with the reaction.

Ehrlich's Test

Aim

To determine the presence of protein in a given sample by Ehrlich's test.

Principle

Ehrlich's test is specific for tryptophan. Ehrlich's test is also known as p-dimethylaminobenzaldehyde (DMAB) and is named after Ehrlich's reagent (Ehrlich's reagent contains p-dimethylaminobenzaldehyde or DMAB, which is used to detect indoles and ubilinogen). Under acidic conditions, DMAB reacts with tryptophan, and DMAB undergoes electrophilic substitution at the indole ring of tryptophan, which results in blue-violet and pink-red coloured condensation products (Fig. **5.5**) [1]. A high concentration of nitrate in a sample can give rise to false results.

Tryptophan + Ehrlich's reagent(DMAB) ⟶ Blue-violet condensed product

Fig. (5.5). Ehrlich's test reaction.

Requirement

❖ ***Chemical***

• Ehrlich's reagent (Take 1.25 gm of DMAB in 50 ml of 95% ethanol. To this, add 50 ml of 10% H_2SO_4).

• Conc. HCl

❖ ***Glassware***

• Test tube

• 1 ml and 5 ml pipette

• Glass dropper

❖ ***Others***

• Test tube holder

• 1 ml and 5 ml pipette

Procedure

1. Take 3 ml of a given sample in a test tube.

2. Add 1 ml of conc. HCl. **(Caution: In this step, you are adding conc. HCl to the aqueous solution, keep hands steady while adding acid).**

3. To that, add a few drops of Ehrlich's reagent and mix it gently.

4. Observe for any colour changes.

Note: Ehrlich's reagent is highly unstable, and it is suggested to prepare freshly before every work.

Results

❖ ***Positive Test***

The formation of violet colour means tryptophan is present. If pink colour formation occurs, then the sample is urobilinogen rich.

❖ ***Negative Test***

No colour formation, which means the absence of tryptophan.

Applications

• Specifically used to detect tryptophan.

• The test is used for the diagnosis of disorders related to the hemolytic process of obstruction of the common bile duct.

• The test can also be implied for the detection of psychoactive compounds and drugs.

• It is majorly used for urobilinogen and bilirubin detection during the diagnosis of liver disease and hemolytic jaundice.

Limitations

• False-negative results can occur in urine samples as nitrates oxidize urobilinogen to urobilin.

• A false negative may also occur during the antibiotic therapy as some of the urobilinogen-producing gut microbes may get destroyed.

Did you know?

Ehrlich's test is mainly used in the clinical diagnosis of liver disease, carcinoid syndrome, and hemolytic processes. And majorly to determine urobilinogen in urine (urobilinogen are bile pigments that can be found in the urine in case of liver disease) or bilirubin (excessive formation during hemolytic jaundice). It is widely used for spot detection of tryptamines (Ex.- DMT) and ergoloids (Ex.- LSD).

ARGININE

Sakaguchi Test

Aim

To determine the presence of protein in a given sample by the Sakaguchi test.

Principle

Arginine, under the alkaline condition, reacts with α- naphthol (1-hydroxy naphthalene) and hypobromite or hypochlorite and produces a red colour complex see Fig. (**5.6**) [1].

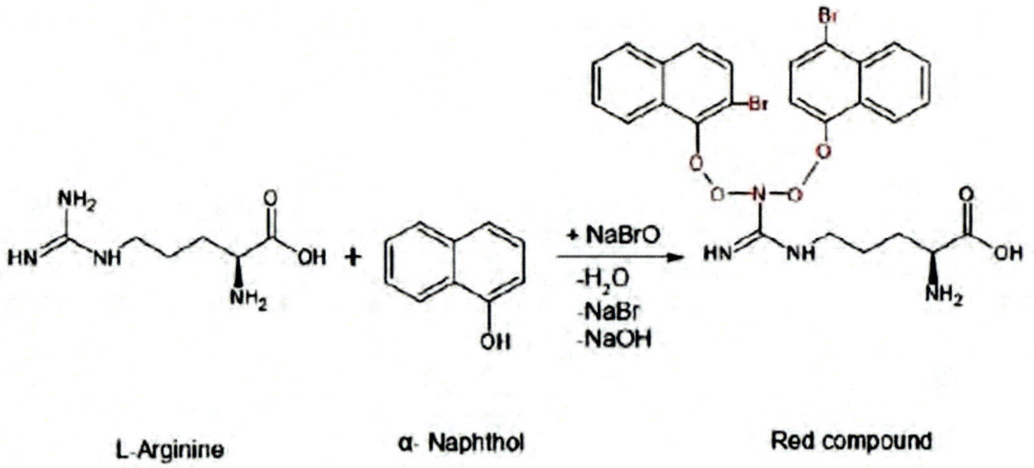

Fig. (5.6). Sakaguchi test reaction

Requirement

❖ ***Chemical***

- Sodium hydroxide (10%W/V)

• α-Naphthol reagent (1%W/V)

• Hypobromite solution (Add 1% of pre-chilled bromine (l) of 5% NaOH and mix it well).

❖ ***Glassware***

• Test tube

• 1 ml pipette

• Glass dropper

❖ ***Others***

• Test tube holder

• Vortex

Procedure

1. Take 1 ml of a prechilled given sample in a test tube and add 0.1-0.2 ml of NaOH and mix it gently.

2. Add 2 drops of alpha naphthol to the solution and mix it gently.

3. Now add 0.2-0.4ml of hypobromite reagent to it and observe for colour change.

Results

❖ ***Positive Test***

The formation of Red colour; means arginine is present.

❖ ***Negative Test***

No colour formation means the absence of arginine.

Application

• Used to detect arginine in free or combined form in protein.

Limitation

• The reaction rate is very slow, thus the formation of colour may take some time.

CYSTEINE

Lead Sulfide Test

Aim

To determine the presence of protein in a given sample by Lead sulfide test.

Principle

Sulphur containing amino acids, such as cysteine and cystine, upon boiling with sodium hydroxide, yields sodium sulfide because of incomplete conversion of the organic sulphur to inorganic sulfide. This inorganic sulfide is identified by converting it into lead sulfide (black precipitate), by lead acetate solution (Fig. **5.7**) [1].

Requirement

❖ ***Chemical***

- Lead acetate solution (10%W/V)
- Sodium Hydroxide (40% W/V)

❖ ***Glassware***

- Test tube
- 1 ml pipette
- Glass dropper

❖ ***Others***

- Test tube holder
- Bunsen burner
- Vortex

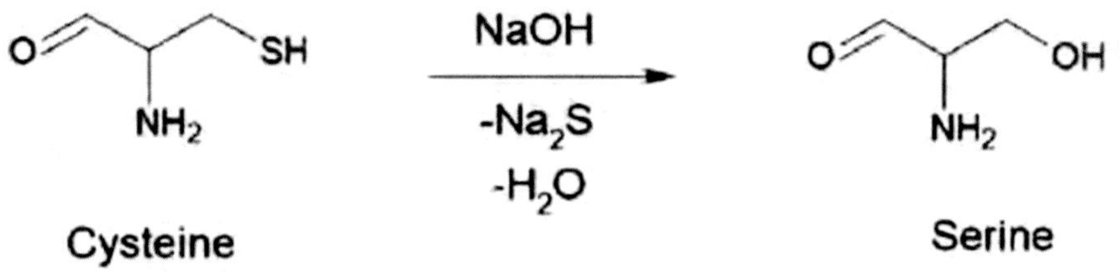

$$Pb(CH_3COOH)_2 + NaOH \longrightarrow Pb(OH)_2 + 2\,CH_3COONa$$

$$Pb(OH)_2 + 2\,NaOH \longrightarrow Pb(ONa)_2 + 2\,H_2O$$

$$Na_2S + H_2O + Pb(ONa)_2 \longrightarrow \underset{\text{Black precipitate}}{PbS\downarrow} + 4\,NaOH$$

Fig. (5.7). Lead sulfide test reaction.

Procedure

1. Take 1 ml of a given sample in a test tube and add 0.1-0.2 ml of 40% NaOH and mix the solution gently.

2. Now boil the solution for 4-9 min using a Bunsen burner.

3. Allow the content to cool down and then add a few drops of 10% lead acetate solution and observe for any colour change.

Results

❖ ***Positive Test***

The formation of black colour precipitation means cystine and cysteine are present.

❖ ***Negative Test***

No black colour precipitation formation means the absence of cystine and cysteine.

Application

• Cystine stones in the kidney and bladder can be identified by this test.

Limitation

• Methionine doesn't respond to this, as sulfur in the thioester bond in Methionine is released by the treatment of NaOH.

• Excess amounts of lead acetate in the reaction mixer may lead to white-coloured precipitation.

Nitroprusside Test

Aim

To determine the presence of protein in a given sample by Nitroprusside test.

Principle

The nitroprusside test aims to detect the -SH group in amino acids or, to be specific, to detect cysteine in amino acids. However, cysteine and methionine both have a -SH group but the -SH group in methionine is involved in thioester linkage, which is difficult to break, so methionine cannot be detected by this test. Under alkaline conditions, the -SH group of cysteine reacts with nitroprusside and produces a red-brown coloured complex (Fig. **5.8**). As the test requires a free -SH group to react with nitroprusside, many proteins do not respond to it but, if the protein sample is heated and thus coagulated and denatured, it facilitates liberation of free -SH group which then can react with nitroprusside and give a positive result [1, 3].

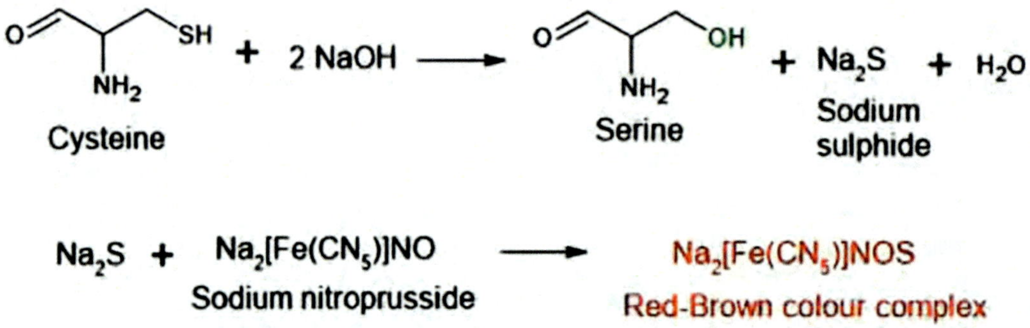

Fig. (5.8). Nitroprusside test reaction.

Requirement

❖ ***Chemical***

- 3M NaOH
- 2% Sodium nitroprusside solution

❖ ***Glassware***

- Test tube
- 2 ml pipette
- Glass dropper

❖ ***Others***

- Test tube holder
- Vortex

Procedure

1. Take 2 ml of 2% cystine and 2% glycine solution in two different test tubes.

2. Add 2 ml of 3M NaOH to both test tubes and give a gentle mixing.

3. Now add 2% sodium nitroprusside solution to both test tubes and mix gently (Precaution- sodium nitroprusside is toxic and must be handled carefully. Inhaling and ingesting the compound must be avoided strictly).

Results

❖ ***Positive Test***

The formation of red colour means cysteine is present.

❖ ***Negative Test***

No colour formation means the absence of cysteine.

Precautions

- Sodium nitroprusside is toxic and must be handled carefully. Inhaling and ingesting the compound must be avoided strictly.

Applications

• As nitroprusside specifically detects cystine, it can be utilised to determine the presence of cysteine in protein samples.

• Nitroprusside test can be used for the diagnosis of cystinuria (the presence of cystine in urine is a pathological indication of cystinuria).

• The test can be implied to detect the presence of ketone in blood and urine for diagnosis of ketonurea and ketonemia.

Limitations

• The above test cannot detect methionine as the -SH group of methionine is not free.

• In some cases, the protein sample may need to be coagulated to get a free -SH group.

Ninhydrin Test

Aim

To determine the presence of protein in a given sample by Ninhydrin test.

Principle

The free amino group and free carboxylic group of amino acids react with ninhydrin (triketohydrindene hydrate) and produce a coloured product named diketo hydrin popularly known as Ruhemann's purple. In some cases, a secondary amino group is attached to amino acids; they also react with ninhydrin. However, when ninhydrin reacts with a secondary amino group it gives yellow colour see Fig. (**5.9**). The primary amines and ammonia react on similar lines with ninhydrin but without the discharge of carbon dioxide [1, 3].

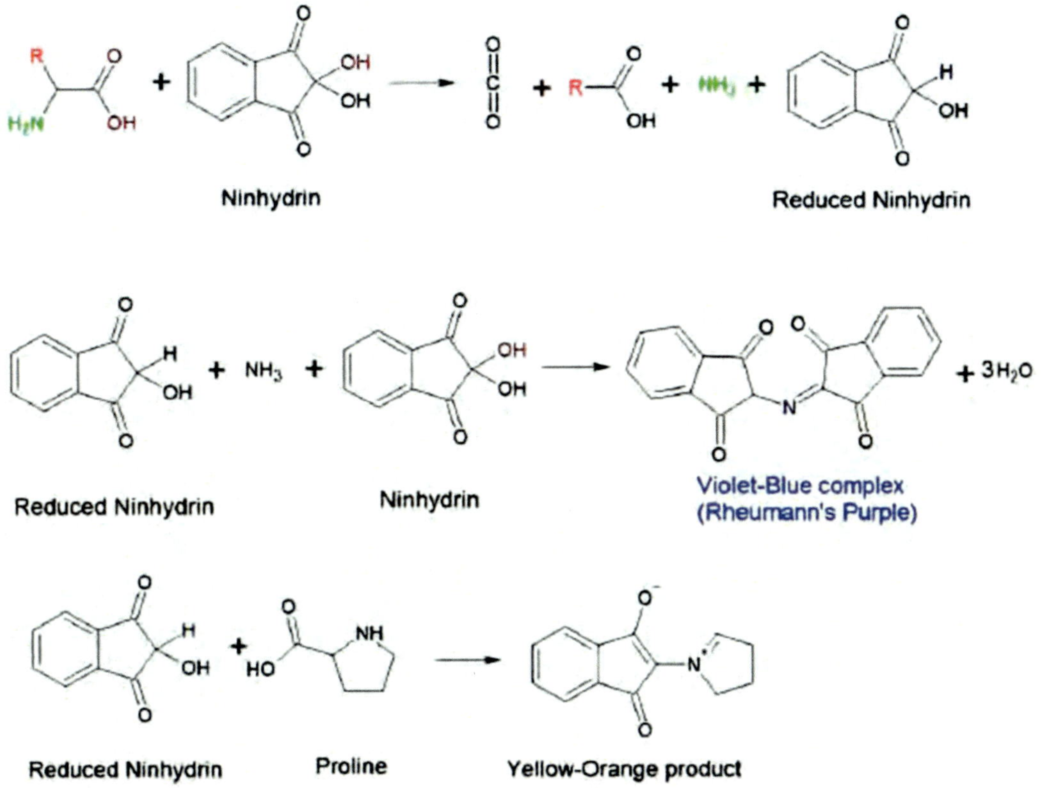

Fig. (5.9). Ninhydrin test reaction.

Requirement

❖ ***Chemical***

Ninhydrin reagent (2% Ninhydrin (w/v) prepared in acetone.)

- Test tube
- 1 ml pipette
- Glass dropper

❖ ***Others***

- Test tube holder
- Boiling water bath
- Vortex

Procedure

1. Take 1 ml of the given sample in a test tube, add 0.1-0.2 ml of ninhydrin reagent and mix it gently.

2. Transfer the test tubes into a boiling water bath for 5 minutes and observe the colour change.

Results

❖ ***Positive Test***

The formation of purple colour; means amino acids are present. If yellow colour forms, then the sample has proline.

❖ ***Negative Test***

No colour formation means the absence of amino acids.

Precaution

• The ninhydrin-acetone reagent is highly flammable and toxic, and causes skin irritation. The reagent must be kept away from the Bunsen burner. Inhaling the reagent must be avoided.

Application

• As this test is quite sensitive, it is commonly used for fingerprint detection (terminal amines of a lysine residue in peptides or proteins shed off in fingerprints which then react with ninhydrin) [2].

Limitation

• It reacts with α amino groups and also with ammonia nitrogen and other free amines.

Did you know?

α-amino acids are molecules that have a carbon molecule at the center and are bounded to an amine group and a carboxylic group.

CONCLUSION

Different tests for qualitative protein determination are discussed in this chapter. These tests are to identify the presence of protein in a given sample. Amino acids present in the protein react with certain chemical compounds to form a colour complex. This colour formation acts as an indication of protein presence.

VIVA VOCE

1. What are α-amino acids?

Ans. α-amino acids are molecules that have a carbon molecule at the center and are bounded to an amine group and a carboxylic group (Fig. **5.10**).

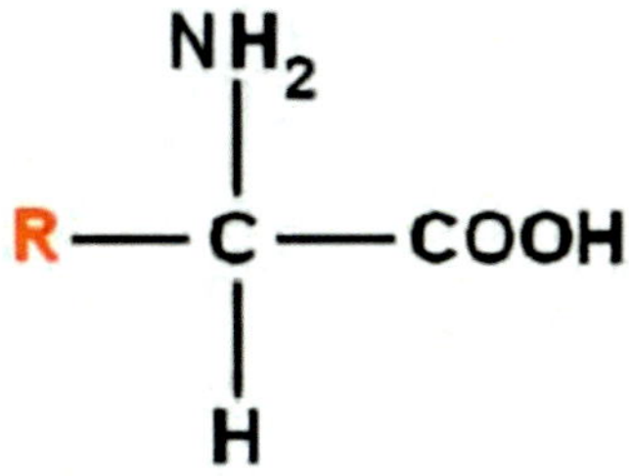

Fig. (5.10). Structure of α-amino acid.

2. What is the role of conc. sulphuric acid in the Hopkin's-Cole test?

Ans. The indole group of tryptophan reacts with acetic acid – glyoxylic acid reagent in the presence of conc. H_2SO_4 produces a purple colour ring.

3. Why do only aromatic amino acids respond to the Xanthoproteic acid test?

Ans. Aromatic amino acids contain an activated benzene ring, which can readily undergo nitration. Conc. HNO_3 nitrates the aromatic phenyl ring and yields a yellow colour complex.

4. Name a few tests to identify tryptophan. In which qualitative test does it form a purple colour?

Ans. Pauly's Diazo test, Histidine test, Hopkin's-Cole test.

In the Hopkin's-Cole test, it forms a purple colour.

5. Predict the amino acids present in a given sample based on the results (results are given below) obtained during qualitative analysis?

Tests	Results
Ninhydrin Test	Positive (Yellow colour)

cont.....

Tests	Results
Xanthoproteic Test	Positive
Millon's Test	Positive
Hopkin's-Cole Test	Negative
Nitroprusside Test	Positive

Ans.

Tests	Results	Possible Amino Acids
Ninhydrin Test	Positive (Yellow colour)	**Proline**
Xanthoproteic Test	Positive	**Tyrosine/Tryptophan**
Millon's Test	Positive	**Tyrosine**
Hopkin's-Cole Test	Negative	**Tryptophan**
Nitroprusside Test	Positive	**Cysteine**

6. Identify the qualitative tests that can be conducted for the given protein sequence?

[Protein sequence -> Ala-Cys-Glu-Ile-Asp-Trp-Val-Tyr-Pro-Ser]

Ans. Based on the protein sequence, the qualitative tests that can be conducted are Nitroprusside test, Millon's test, Xanthoproteic Test, Hopkin's-Cole Test, Lead sulphide test, Pauly's Diazo test, and Ninhydrin test.

7. Give examples of sulphur-containing amino acids?

Ans. Sulphur-containing amino acids are cysteine, Cystine, and Methionine.

REFERENCES

[1] Tiwari, A. *Practical Biochemistry: A Student Companion*; LAP LAMBERT Academic Publishing: United Kingdom Str., **2015**.

[2] Lennard, C. *Forensic sciences: fingerprint techniques*; Reference Module in Chemistry, Molecular Sciences and Chemical Engineering, **2013**, pp. 1-0.

[3] Milio, F.R.; Loffredo, W.M. Qualitative testing of amino acids and proteins. chemical education resources. Inc *(USA). REAC,* **1995**, *448*, 1-2.

CHAPTER 6

Protein Determination by SDS-PAGE

Sai Joshi[1,*]

[1] *CSIR-NCL, Dr. Homi Bhabha Road, Pune - 411 008, India*

Abstract: SDS-PAGE is the abbreviated form of sodium-dodecyl sulphate-polyacrylamide gel electrophoresis. It is a commonly used gel electrophoresis for the qualitative estimation of proteins. It is based on the principle that treatment of protein with SDS, an anionic detergent, causes a constant negative charge throughout the protein surface. It also breaks the protein chain into smaller peptide chains. Hence, the separation of proteins is solely based on the size and shape of the fragments generated and is independent of the charge of the protein. This, combined with polyacrylamide gel electrophoresis, allows for visualization of these fragments where the smaller fragments retain at the bottom of the gel. This method has been discussed in detail in this chapter.

Keywords: Electrophoresis, SDS-PAGE, Size based separation, Protein.

INTRODUCTION

Electrophoresis through agarose or polyacrylamide gels is a standard method used to separate, identify and purify biopolymers, since both these gels are porous in nature. Polyacrylamide gel electrophoresis (PAGE) is a technique widely used in biochemistry, forensic chemistry, genetics, molecular biology and biotechnology to separate biological macromolecules, usually proteins or nucleic acids, according to their electrophoretic mobility. The most commonly used form of polyacrylamide gel electrophoresis is the Sodium dodecyl sulphate-Polyacrylamide gel electrophoresis (SDS- PAGE), used mostly for the separation of proteins which is an analytical method used to separate components of a protein based on size.

PRINCIPLE

SDS-PAGE stands for Sodium Dodecyl Sulphate-Polyacrylamide Gel Electrophoresis. Protein separation is generally achieved using a polyacrylamide

* **Corresponding author Sai Joshi:** CSIR-NCL, Dr. Homi Bhabha Road, Pune - 411 008, India; E-mail: saijoshi1997@gmail.com

Pamela Jha (Ed.)

gel. This gel is prepared by incorporating acrylamide with a polymerizing agent - ammonium persulfate (APS). TEMED (N,N,N',N'-tetramethylenediamine) is involved in catalyzing this polymerization reaction by enhancing the production of free radicals by APS. This converts acrylamide to polyacrylamide. The sievelike structure of polyacrylamide is manipulated to achieve protein separation [1 - 5]. The protein separation takes place based on the pore size present in the polyacrylamide gel, as shown in Fig. (**6.1**). The stacking gel allows all the protein samples to line up along the same length in the gel so that they enter the resolving gel at the same time. The resolving gel causes the separation of proteins based on molecular weight.

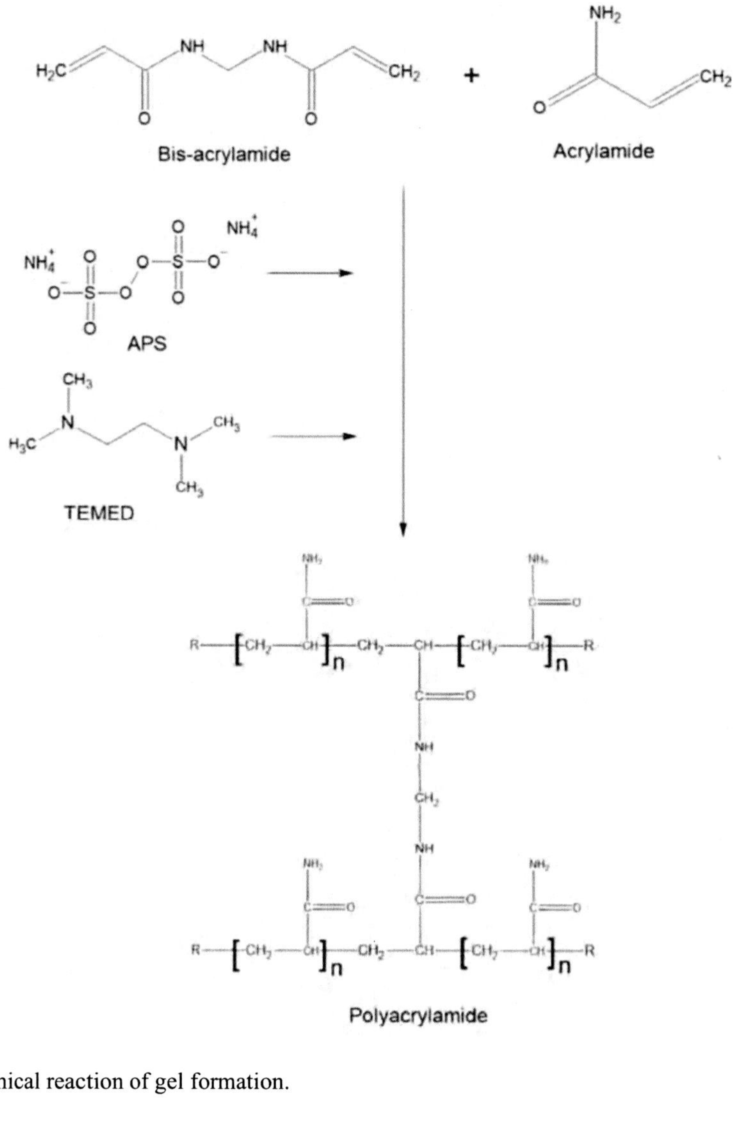

Fig. (6.1). Chemical reaction of gel formation.

In this technique, primarily the protein sample is treated with DTT (dithiothreitol) or 2-mercaptoethanol, as shown in Fig. (**6.2**). This breaks the disulphide bonds created by cysteine and methionine in the amino acid chain of the protein [1 - 3] [5]. Consequently, the binding of SDS to the protein sample denatures the protein and covers the protein sample with an overall negative charge. At pH 7, in the presence of 1% w/v SDS and 2-mercaptoethanol, proteins dissociate into their subunits and bind large quantities of detergent. Under these conditions, most proteins bind about 1.4g of SDS per gram of protein which completely masks the natural charge of proteins giving a constant charge mass to ratio. Hence, the larger the molecule, the greater will be the charge; this can be detected according to the electrophoretic mobility of the complex. Hence, the proteins get separated based on molecular weight see Fig. (**6.2**) [1 - 5].

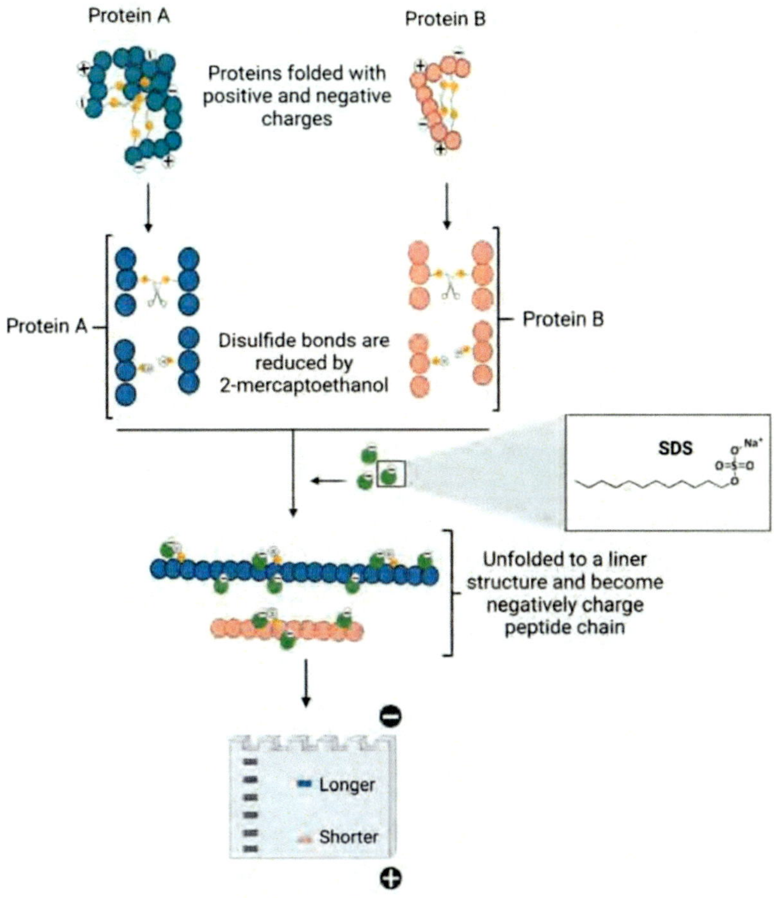

Fig. (6.2). Protein denaturation.

REAGENT PREPARATION

1. 1.5 M Tris-HCl (pH-8.8) (For 10 ml)

Weigh 2.36 gm of Tris base and dissolve it in 5 ml distilled water. Maintain the pH to 8.8 and then add distilled water to make the total volume 10 ml.

2. 0.5 M Tris-HCl (pH-6.8) (For 10 ml)

Weigh 0.78 gm of Tris base and dissolve it in 5 ml distilled water. Maintain the pH to 6.8 and then add distilled water to make the total volume 10 ml.

3. 10% SDS buffer

Weigh 10 gm of SDS and dissolve it in 20 ml distilled water. Then add distilled water to make the total volume 100 ml.

4. 10% APS (Freshly prepared)

Weigh 0.1 gm of APS and dissolve it in 1 ml distilled water in an Eppendorf tube.

5. TEMED (5% v/v)

6. Acrylamide solution (30%) (For 10 ml)

Acrylamide 3.857 gm

Bis-acrylamide 0.133 gm

Water 10 ml

Filter and store the solution in a dark bottle.

Caution: the monomer is highly toxic, so avoid inhalation and skin contact.

REQUIREMENTS AND PROCEDURES

Preparation of Gel

Requirements

(a) 70% ethanol

(b) PAGE apparatus

(c) Distilled water

(d) Agar

(e) Resolving gel (12%) [For 10 ml]

• Water 3.4 ml

• 1.5M Tris 2.5 ml

• 10% SDS 0.1ml

• 30% Acrylamide 4 ml

• 10% APS 50 µl

• TEMED 5 µl

(f) 1.5 M Tris-Cl (pH 8.8)

(g) Acrylamide

(h) SDS

(i) Ammonium persulfate

(j) TEMED

(k) Stacking gel (5%)

• ddH2O 6.1 ml

• 0.5M Tris 2.5 ml

• 10% SDS 0.1 ml

• 30% Acrylamide 1.3 ml

• 10% APS 50µl

• TEMED 10µl

(l) Miscellaneous: Beakers, Pipettes,

Procedure

1. Clean the glass plates with ethanol. Assemble the glass plates and spacers on the SDS apparatus.

2. Assemble the glass plates by putting the shorter glass plate in front of the longer one.

3. Fix the glass plates and fill them with distilled water to check for any leakage from the bottom. If leakage occurs, seal the end of the plate with molten agar.

4. First, the resolving gel is prepared, which is usually more basic and has higher polyacrylamide content than the loading gel, using the given composition.

5. Take a small beaker, and add distilled water, Tris-Cl (pH-8.8), acrylamide, and SDS with the help of a pipette.

6. APS and TEMED are added when the gel is ready to be polymerized.

7. Mix it properly by moving the beaker in a circular motion on the surface.

8. Pour the gel between the plates before it polymerizes, then add water to the surface to remove the unpolymerized gel.

9. Similarly, prepare to stack gel using the following composition.

10. When resolving gel polymerizes, remove the water by using tissue paper,

11. Pour the stacking gel solution and for creating the wells, place the comb.

12. After the stacking gel polymerizes, remove the comb.

13. The prepared polyacrylamide gel is ready for electrophoresis.

Preparation of Samples

Requirements

(a) 2X Sample buffer

• ddH2O 125µl

• 0.5 M Tris, pH (6.8) 125 µl

• 50% Glycerol 300 µl

• 10% SDS 400 µl

• 2-mercaptoethanol 50 µl (add immediately before use)

• Pinch of Bromophenol Blue

(b) Miscellaneous: Conical flask, beaker, pipette

Procedure

1. Mix the protein sample with an equal volume of 2X sample buffer.

Loading the Sample

Requirements

(a) Running buffer

• Tris 3.002g

• Glycine 14.40 g

• SDS 1 g

(b) Standard marker

(c) Miscellaneous: weighing machine, micropipette and tips.

Procedure

1. Clamp the gel plates properly and fill the buffer chamber with a running buffer.

2. Load 22 μL of prepared samples into the wells.

3. The first marker is loaded in the first lane to determine the size of the desired protein.

4. Then, samples are loaded in the adjacent lanes. Bubbles are avoided in the tip.

Running the Gel

Requirements

(a) Voltage supply

Procedure

1. The lid is placed on top of the buffer chamber.

2. Connect the electrical leads to the power pack with the proper polarity (black to black and red to red) and run the gel at 60-100V.

3. Run the gel till the dye reaches the end of the gel.

4. As soon as the dye reaches the end, switch off the power.

Staining the Gel

Part A - Coomassie Blue Staining

Requirements

(a) Spatula

(b) Staining solution

- Coomassie blue 0.2% w/v
- Methanol: Acetic acid: water 5:1:5
- Dissolve Coomassie Blue in Methanol: Acetic acid: Water.
- Make up the final volume to 500 ml

Procedure

1. Take out the glass plates from the electrophoretic tank and tear apart the two plates using a spatula.

2. Carefully place the gel into a staining solution.

3. Incubate the gel into this staining solution overnight.

Note: The staining of gel is always done in a covered container.

Part B - Silver Staining

Requirements

(a) Fixation solution

- 40% ethanol
- 10% acetic acid
- 50% water

(b) Protein treatment solution

• 20% ethanol

• 5% acetic acid

• 75% water

• 4mg dithiothreitol

(c) 0.5% dichromate

(d) Water

(e) 0.1% silver nitrate

(f) Complex formation solution (pH 12)

• 0.02% paraformaldehyde

• 3% sodium carbonate

(g) 1% acetic acid

Procedure

1. Place the gel in the fixation solution for 30 minutes.

2. Remove the fixation solution.

3. Place the gel in the protein treatment solution for 30 minutes.

4. Remove the protein treatment solution.

5. Wash the gel with 0.5% dichromate for 5 minutes and then with water for 5 minutes.

6. Wash the gel with water for 5 minutes.

7. Equilibrate the gel with 0.1% silver nitrate for 30 minutes.

8. Then, wash the gel with water for 1 minute.

9. Incubate the gel in a complex formation solution.

10. Add 1% acetic acid to stop the complex formation.

Destaining the Gel

Requirements

(a) Destaining solution

Procedure

1. Once the gel is stained, the gel is then transferred from a staining solution to a destaining solution.

2. Destaining is done until all the extra dye is removed from the gel except the dye bound to protein in the form of bands.

Visualization of the Gel

- Visualize the gel patterns using a gel doc system.

- After destaining, clear bands of protein are seen and compared to the corresponding bands' ladder to determine the molecular weight of the protein (Table **6.1**).

Table 6.1. Representative Gel images with the comments.

Gel Image	Comments
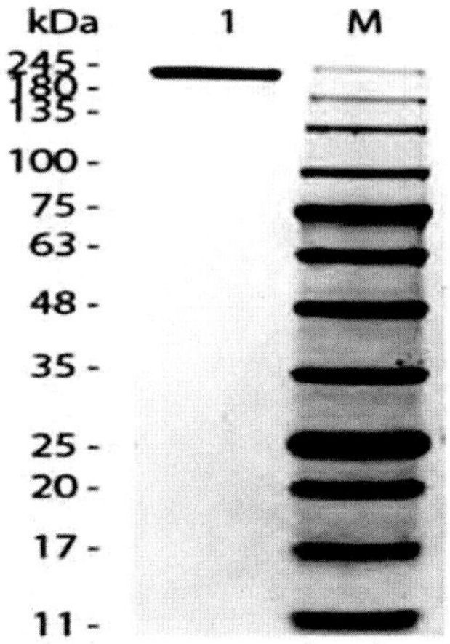	• This gel image shows a proper separation of proteins. • The ladder has separated properly, hence preparation of the gel or voltage supply was proper. • The bands are distinct. **Interpretation:** Lane 1: Sample 1 has a protein of molecular weight of 245KDa. Lane 2: Ladder (Mol. weight 11KDa-245KDa).

(Table 6.1) cont.....

Gel Image	Comments
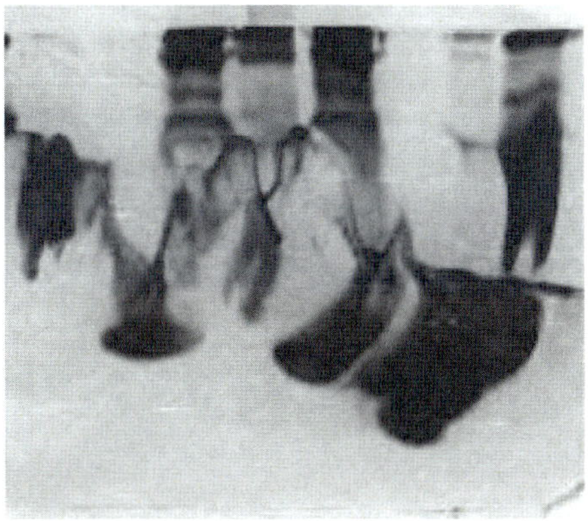	• The pouring of the gel was not proper; hence the lower half of the gel has begun to polymerize but the upper part has not, because of this, a smeared gel appeared. • Hence, this is an example of a smeared protein gel. **Interpretation:** Smearing can have a variety of causes, but most commonly it is due to an unevenly poured acrylamide mixture or due to gross overloading of protein.
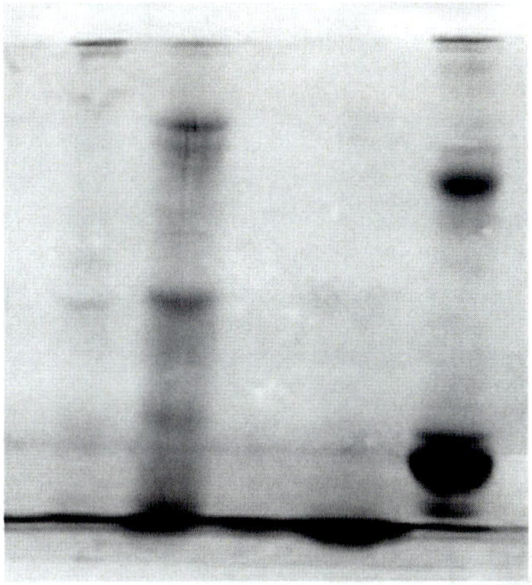	• Due to the presence of the background as dark and uneven, it can be interpreted that the protein was in the running buffer itself. • The apparatus was disturbed, because of which the samples were slopped out of the wells. • It is also probable that this occurred due to an uneven voltage supply across the gel. **Interpretation:** The spilt sample would not resolve into bands, since it continuously entered the gel from the upper buffer compartment.
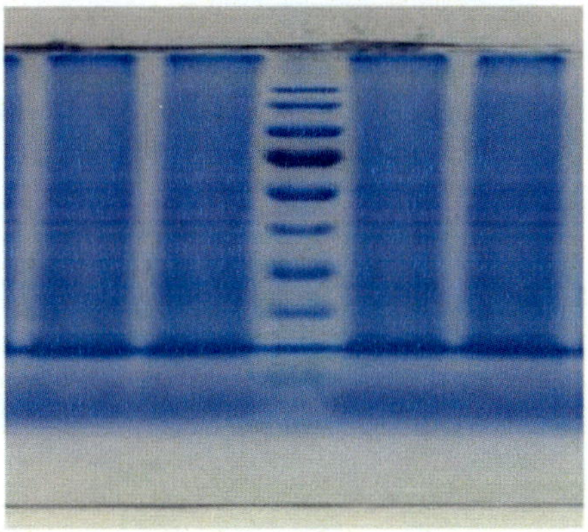	• No protein bands are observed. All the lanes show background staining. **Interpretation:** De-staining is to be performed for a longer time to reduce background staining due to which protein bands are not visible. If that doesn't remove the stain, the sample may contain interfering compounds.

(Table 6.1) cont.....

Gel Image	Comments
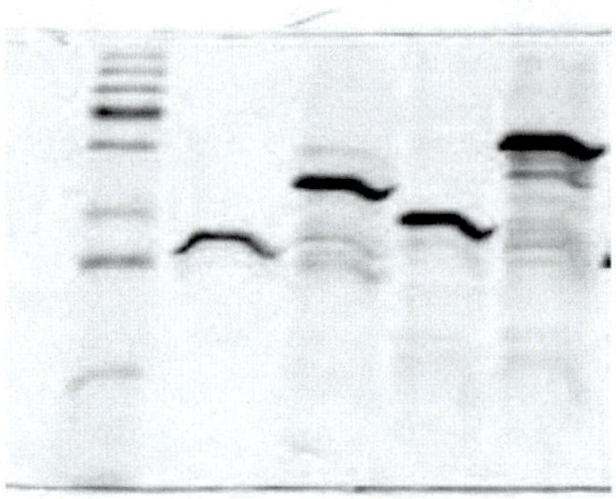	• Frowning bands are observed. That is the curve pattern facing downwards at the ends. **Interpretation:** This suggests that either protein has been overloaded, sample preparation error (try to minimize slats), or incorrect running conditions concerning the voltage.
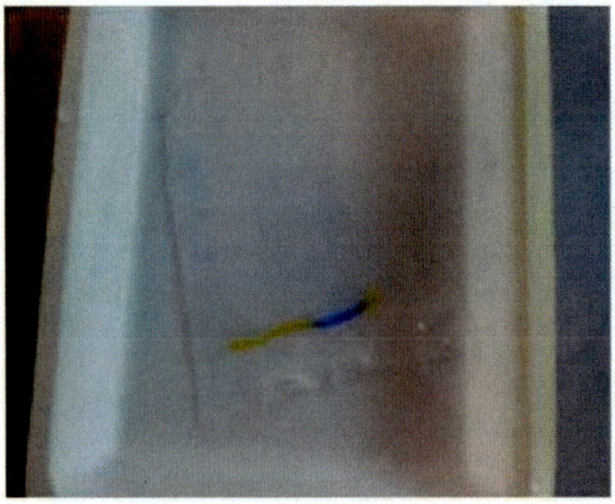	• A yellow band is observed. **Interpretation:** Here, if the loading dye is bromophenol blue, we can predict that the sample buffer is too acidic due to the yellow band observed. Add Tris base until the buffer turns bl ue again.
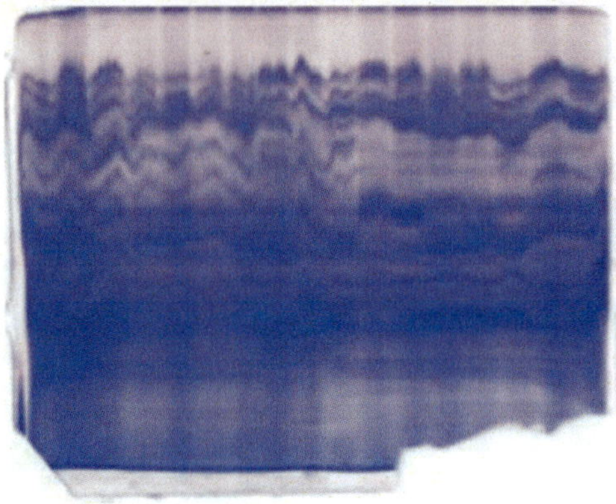	• Very uneven bands are observed with constant fluctuations. **Interpretation:** It was run with a high current and overheated. Try running the gel again with a low current.

PRECAUTION

1. Ensure that the gel keeps immersed in the buffer solution, if not, it might result in a cracked gel.

2. SDS is a neurotoxin. Hence, it can be harmful to the nervous tissue. It can act as an irritant to the eyes and skin. It is always necessary to wear gloves and eye goggles.

3. 2-mercaptoethanol is poisonous and harmful. It also has a penetrating smell, avoid inhaling or ingesting it.

APPLICATIONS

1. To analyze the molecular weight of protein samples.

2. To perform peptide mapping.

3. To estimate the purity of the protein sample.

4. To quantify proteins.

5. It is a necessary step to perform western blotting.

Did you know?

Sample loading buffer is also known as the Laemmli buffer after the Swiss professor, Ulrich Laemmli, in the 1970s.

CONCLUSION

Sodium Dodecyl Sulphate-Polyacrylamide Gel Electrophoresis or SDS-PAGE is a technique where the protein molecules are denatured first and then run on the SDS gel. Based on the size and charge of the denatured protein molecules, they are separated on the resolving gel. This is most common and routinely practiced in the lab. Through this, we can also quantify the size of the protein molecule, and further analysis can be done on the separated protein molecules.

VIVA VOCE

1. How is SDS-PAGE different from Native-PAGE?

Ans. SDS PAGE denatures the protein sample and creates a uniform negative charge on the protein sample; due to this reason, the proteins are separated only by size and shape. However, Native-PAGE preserves structure in the active form, and the separation observed is due to the size and shape as well as the charge of the protein.

2. How many bands of IgG will be formed by SDS-PAGE under reducing conditions?

Ans. There will be 2 bands on an SDS-PAGE gel, one of heavy chain and another of light chain, corresponding to 50KDa and 25KDa, respectively. This is observed as the total size of the protein is 150KDa with 2 heavy and 2 light chains.

3. Why are proteins not separated based on charge in SDS-PAGE?

Ans. In SDS-PAGE, the anionic detergent, SDS completely masks the natural charge of proteins giving a constant negative charge to the proteins. Hence, separation based on the charge is not possible.

4. How do we decide if Coomassie blue staining or silver staining is required in SDS-PAGE?

Ans. In SDS-PAGE, Coomassie blue staining is usually used because it is simple and easy to use. However, it can only detect proteins which are present in the micrograms range (0.3-1μg/protein band). For a more sensitive staining, silver staining is used, which can detect nanograms of proteins as well (2-5ng/band of protein).

REFERENCES

[1] Berkelman, T. Quantitation of protein in samples prepared for 2-D electrophoresis. *2D PAGE: Sample Preparation and Fractionation,* **2008**, 43-49.

[2] Fifield, R. Book Review: Electrophoresis: Theory, techniques and biochemical and clinical applications. *Ann. Clin. Biochem: Inter. J Laboratory Med.,* **1987**, *24*(3).

[3] Rabilloud, T. Variations on a theme: Changes to electrophoretic separations that can make a difference. *J. Proteomics,* **2010**, *73*(8), 1562-1572. [http://dx.doi.org/10.1016/j.jprot.2010.04.001] [PMID: 20394846]

[4] Steinberg, T.H. Protein gel staining methods: An introduction and overview. *Methods Enzymol.,* **2009**, *463*, 541-563. [http://dx.doi.org/10.1016/S0076-6879(09)63031-7] [PMID: 19892191]

[5] Hames, B.D., Ed. *Gel electrophoresis of proteins: A practical approach*; OUP Oxford, **1998**.

CHAPTER 7

Protein Determination by Western Blotting or Immunoblot

Sinchan Hait[1,*]

[1] *Department of Chemical Engineering, BITS-Pilani K.K. Birla Goa Campus, Goa-403726, India*

Abstract: Western blotting, also known as immunoblot, is a chromatography-based technique in which quantitative detection of proteins can be performed based on their molecular weight. This is often performed after the SDS-PAGE technique. It is used to identify, analyze and quantify proteins from a protein mixture. The major steps include 1. Separation of proteins (*e.g.*, SDS-PAGE), 2. Transfer onto a solid support, 3. Utilizing primary and secondary antibodies for visualization. Western blotting can be used qualitatively and quantitatively. In this chapter, both of these techniques are explained in detail.

Keywords: Chromatography, Immunoblot, Protein, SDS-PAGE.

INTRODUCTION

Western blotting is a technique broadly used in research to separate and identify proteins. The separation of proteins based on size can be performed using SDS-PAGE; the size-based separated bands are then transferred onto a solid support such as a nitrocellulose membrane, which allows for visualization of the band for each protein. The incubation of the primary antibody with the membrane allows for the interaction between the protein of interest and the antibody specific to that protein. The bound antibody is detected through a secondary antibody which acts as a chromogen system (*e.g.*, HRP-conjugated protein) [1 - 5]. This technique can be used for both qualitative and quantitative analysis of protein samples. In this chapter, both techniques are discussed.

* **Corresponding author Sinchan Hait:** Department of Chemical Engineering, BITS-Pilani K.K. Birla Goa Campus, Goa-403726, India; E-mail: sinchan.hait1@gmail.com

QUALITATIVE WESTERN BLOTTING

Principle

The principle of western blotting (WB) or immunoblot is based on immuno-chromatography. The protein molecules are separated into polyacrylamide based on their molecular weight. Once proteins are separated on polyacrylamide, they are electro-transferred onto a nitrocellulose membrane or polyvinylidene difluoride (PVDF) [1 - 4]. Proteins attach to the membrane by hydrophobic interaction. Then these isolated proteins are detected by conjugation with a specific primary antibody and enzyme-labelled secondary antibody and substrate (Table **7.1**) by autoradiography or fluorescence or chemiluminescent methods.

Table 7.1. Colorimetric enzyme and substrate system for antigen detection of western blotting.

Enzyme	Substrate	Colour
Alkaline phosphatase	5-Bromo-4-chloro-3-indolyl phosphate (BCIP) with Nitro blue tetrazolium chloride (NBT)	Black or purple
Horseradish peroxidase	Naphthol AS-MX phosphate with fast red TR	Bright red
	4-Chloro-1-naphthol (CN)	Blue or purple
	3-Amino-6-ethyl carbazole (ACE)	Red or brown
	3,3'-Diaminobenzidine tetrahydrochloride (DAB) [carcinogenic]	Brown
Glucose oxidase	Phenazine methosulfate with NBT	Black or purple
β-Galactosidase	X-Gal	Blue

Did you know?

W. Neal Burnette discovered the western blotting technique taking inspiration from the northern blotting technique, but in 1979 his manuscript was rejected, and finally, in 1981, his manuscript was accepted and published in Analytical Biochemistry [4].

Requirement

❖ 1X transfer buffer

• 25 mM Tris base

• 192 mM glycine

• 0.1% SDS

• Adjust pH to 8.3

❖ Filter paper towel

❖ 5% skim milk

❖ 5% BSA

❖ Polyvinylidene fluoride (PVDF) membrane

❖ SDS-PAGE apparatus

Procedure

SDS-PAGE

See Chapter 6 for the SDS-PAGE procedure.

Electrotransfer

1. Cut one polyvinylidene fluoride (PVDF) membrane with the same dimensions as the gel.

2. Wet the sponge and filter paper in the transfer buffer and wet the PVDF membrane in methanol.

3. Preparation of the sandwich for electrotransfer is as follows: Sponge, 3 Filter Papers, Gel, PVDF, and 3 Filter Papers (Caution: ensure no air bubble present).

4. Keep the sandwich in the transfer apparatus, place it on ice, and the temperature during the process must be maintained at 4°C.

5. Pour the transfer buffer into the apparatus, and ensure that the sandwich is covered with the buffer.

6. Connect the apparatus to the electrical supply (caution: ensuring that the PVDF membrane is between the gel and a positive electrode).

7. Transfer for 90 minutes (Tip: running time depends upon gel thickness).

Blocking and Antibody Incubation

1. Take out the PVDF membrane and dip it in 5% skim milk in a buffer for 1 hour.

2. Remove the PVDF membrane and add primary antibodies and keep it in 5% BSA and incubate at 4°C on a shaker.

3. After incubation, wash the membrane with TBST for 5 min and repeat the step 3 times.

4. Add enzyme-conjugated secondary antibodies in 5% BSA and incubate at 4°C on a shaker for 1 hour.

5. After incubation, wash the membrane with TBST 3 times and for 5 mins each.

Detection

1. Prepare ECL to mix. Incubate the membrane for 1-2 minutes (Tip 1: for better results, use a 100 μL pipette to ensure ECl covers the whole surface of the membrane, Tip 2: for sensitive substrate, keep shorter incubation time or even dilution to achieve an optimal result).

2. Carefully remove the membrane.

3. Expose the membrane to autoradiography film or detect it with chemiluminescent detection systems and visualize the result in the dark room (Tip: in the beginning, use multiple exposure lights to identify optimal exposure time and an initial 10-sec exposure will provide exposure time).

QUANTITATIVE WESTERN BLOTTING

Principle

Western blotting can also be modified to quantify the given protein sample. In quantitative western blotting, the accepted protein sample load for separation is 10μg to 100μg. The protein sample is normalized first, and internal loading must be used during the process. If the protein sample is not examined, then saturation may go undetected [5]. Saturation may occur as:

• Membrane saturation: It occurs when a sample is overloaded. To alleviate this, a standard bond density Vs. protein concentration needs to be prepared for each sample by ½ dilutions starting from 100μg load over a minimum of 12 dilutions on a stain-free gel, which then can be detected by a stain-free detection system.

• Signal saturation: It can be caused by an intensive band that exceeds the detection ability of the imaging system. In a saturation signal, the signal becomes a plateau.

Internal loading control (ILC) is an integral part of qualitative WB, these are housekeeping proteins (HKP) such as β-actin, tubulin (s), and glyceraldehyde-3-phosphate dehydrogenase (GAPDH) [5]. But the serial dilution samples must be used to identify appropriate HKP. Normalization is to mathematically correct small, inevitable variations during sample loading and electro-transfer by comparing the target protein to the ILC of each lance and the target protein is normalized by the signal value generated by ILC in that lane. The assumption made in normalization is that ILC and target signal both depend upon sample concentration. The bond intensity of the target protein can be compared to determine if there is any difference between the samples [5].

Interpretation of Test Results

Positive Result

A clear line in the control zone and the test area on the membrane.

Negative Result

A single line in the control zone.

Invalid

A single line in the test area without a corresponding control line.

Precautions

• The membrane on which the protein will be transferred is to be blacked by BSA or non-fat milk powder then, there won't be any non-specific binding of antibodies.

• The membrane must be chosen based on the molecular weight of the protein; for lower molecular weight proteins, nitrocellulose is preferred over PVDF and *vice-versa*. PVDF shows a higher binding ability for high molecular weight proteins.

• For efficient transfer and high-resolution, bis-acrylamide must be used in less quantity.

Troubleshooting

Unexpected Bands

The appearance of unexpected bands can be due to the degradation of protein by protease. In such cases, it is suggested to use fresh samples or alter the antibody. If at a spot, the protein concentration seems to be at high concentrations, then heating the sample can help to break the quaternary structure of the protein.

Blurred Bands

This may be due to high voltage and or may be due to the presence of air bubbles during the electrophoretic transfer. In such situations, the voltage must be kept lower and needs to ensure that no air bubbles are present during the sandwich preparation. In some cases, changing the buffer might help.

No Bands

In some cases, no bands can be due to the non-specific antibody or buffer used. If either of the primary or secondary antibody is not specified, then this issue may arise. Along with that, antibody concentration must be checked, if the concentration is too low, then the signal will be very low. If the exposure time of antibodies is increased, it might help to make the band more visible. Another reason can be the absence or lowest concentration of the antigen. Buffers also can contribute to this problem and hence must be prepared fresh and need to ensure they are free from contamination. If the buffer is contaminated with sodium azide, this can inactivate HRP. Also, non-fat dry milk powder masks antigens. In this case, a decreased amount of milk must be used or use BSA.

High Background

A major reason behind the high background is the high concentration of antibodies. Another issue could be due to the use of an old buffer. An increase in washing time can solve this problem. Exposure time can also lead to a high background, so different exposure times must be checked for the optimum time.

Uneven Bands on Blot

Uneven bands can be caused by improper transfer. During sandwich preparation, if any air bubbles remain, then it will appear as a dark band. It is suggested to use a shaker during incubation. Washing is of utmost importance to overcome the problem of preventing antibodies from binding to blocking agents. Filtering the blocking agents can help to remove contaminants. Another problem is caused due to aggregation of antibodies; in such situations, the secondary antibody must be centrifuged and filtered to remove aggregates.

CONCLUSION

Western blotting, also popularly known as immunoblotting, is an important technique for the identification of proteins in an unknown sample. As discussed, this is often used as a continuation of SDS-PAGE. This technique can be utilized qualitatively and quantitatively. This helps us to better understand the samples.

VIVA VOCE

1. What is Western Blotting?

Ans. Western blotting refers to the electrophoretic transfer of protein samples from SDS-polyacrylamide gel to a nitrocellulose membrane, followed by immune detection attached to fluorescence or chemiluminescent detection.

2. What are the other terminologies used for Western Blotting?

Ans. Immunoblotting and protein blotting.

3. State the pros and cons of using PVDF and nitrocellulose membranes?

Ans.

S. No.	Criteria	PVDF	Nitrocellulose
1.	Pros	• Higher binding capacity • Stronger and resilient • Easy to handle	• Lower background than PVDF • Easier to block • Cheaper compared to PVDF • Better for lower molecular weight proteins
2.	Cons	• Higher background • Expensive • Activated through methanol	• Lowest binding capacity • Fragile- Can be easily cracked

4. Why do we need to transfer proteins from the gel to the membrane?

Ans. The reasons for transferring proteins from gel to membrane are as follows:

A. Membranes are easy to handle and easy to manipulate, whereas gels are fragile and so difficult to handle.

B. Detection of a protein in the gel is difficult. But on the other hand, after transferring them to the membrane, they can bind to antibodies easily (as they are more open for antibodies to bind).

REFERENCES

[1] Gwozdz, T.; Dorey, K. Chapter 6 - Western Blot. In: *Basic Science Methods for Clinical Researchers*; Morteza, J.; Francesca, Y.L.; Saldanha, M.J., Eds.; Academic Press, **2017**; pp. 99-117.

[2] Yang, P-C.; Mahmood, T. Western blot: Technique, theory, and trouble shooting. *N. Am. J. Med. Sci.*, **2012**, *4*(9), 429-434. [http://dx.doi.org/10.4103/1947-2714.100998] [PMID: 23050259]

[3] Xiang, Y.; Zheng, Y.; Liu, S.; Liu, G.; Li, Z.; Dong, W. Comparison of the sensitivity of western

blotting between PVDF and NC membranes. *Sci. Rep.,* **2021**, *11*(1), 12022.
[http://dx.doi.org/10.1038/s41598-021-91521-8] [PMID: 34103620]

[4] Burnette, W.N. Western blotting: Electrophoretic transfer of proteins from sodium dodecyl sulfate-polyacrylamide gels to unmodified nitrocellulose and radiographic detection with antibody and radioiodinated protein A. *Anal. Biochem.,* **1981**, *112*(2), 195-203.
[http://dx.doi.org/10.1016/0003-2697(81)90281-5] [PMID: 6266278]

[5] Taylor, S.C.; Posch, A. The design of a quantitative western blot experiment. *BioMed Res. Int.,* **2014**, *2014*, 1-8.
[http://dx.doi.org/10.1155/2014/361590] [PMID: 24738055]

CHAPTER 8

Quantitative Tests for Protein Detection

Sinchan Hait[1], **Sai Joshi**[2,*] and **Ankita Nitin Padhye**[3]

[1] *Department of Chemical Engineering, BITS-Pilani K.K. Birla Goa Campus, Goa-403726, India*

[2] *CSIR-NCL, Dr. Homi Bhabha Road, Pune - 411 008, India*

[3] *Food Engineering and Technology Department, Institute of Chemical Technology, Mumbai, Maharashtra, India*

Abstract: Quantitative tests for protein are essential for understanding the amount of protein present in a sample. Various tests have been developed, such as BCA, biuret, Bradford, and Lowry tests. These are the most common and standardised procedures for the determination of protein. The mentioned techniques are colourimetric assays. However, over a while, several more advanced techniques are also being developed. In this chapter, quantitative methods for protein are discussed.

Keywords: BCA, Biuret, Bradford, Lowry, Protein.

INTRODUCTION

Quantitative tests help us to understand the concentration of protein in a given sample. For these, BCA, Biuret, Bradford, and Lowry tests are some standard assays which determine the same. These methods are colourimetric assays and the colour formation is equivalent to the concentration of protein resent in it.

BICINCHONINIC ACID (BCA) TEST

Aim

Quantification of protein using Bicinchoninic acid (BCA).

Principle

Bicinchoninic acid, sodium salt, is a stable, water-soluble compound capable of forming an intense purple complex with cuprous ions ($Cu^{'+}$) in an alkaline environment. This reagent forms the basis of an analytical method capable of monitoring cuprous ions produced in the reaction of protein with alkaline Cu^{2+}

* **Corresponding author Sai Joshi:** CSIR-NCL, Dr. Homi Bhabha Road, Pune - 411 008, India;
E-mail: saijoshi1997@gmail.com

Pamela Jha (Ed.)

(biuret reaction). The colour produced from this reaction is stable and increases in a proportional fashion over a broad range of increasing protein concentrations [1].

At room temperature, cysteine, cystine, tryptophan, and tyrosine residues reduce Cu^{2+} to Cu^{1+}. At higher temperatures (37° C to 60° C), peptide bonds in protein also perform this reduction, which produces results that correlate more strongly to protein quantity than to the presence of specific residues. Each Cu^{1+} ion forms a complex with two BCA molecules, creating a purple-coloured product that absorbs light at 562 nm (Fig. **8.1**) [1]. Absorbance data from unknown samples can be plotted against a standard curve and the concentration of unknown samples can thus be calculated.

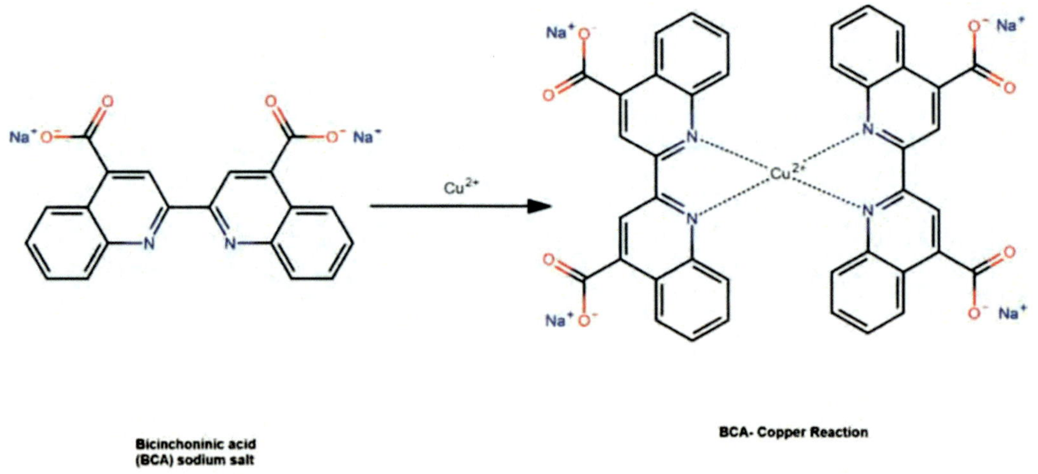

Fig. (8.1). Formation of the purple complex with BCA and cuprousion generated from the biuret reaction.

The bicinchoninic acid (BCA) or copper-based assay (range: 20-2000 ug/ml).

Requirements

• Reagents: Standard Bicinchoninic acid, sodium salt, Folin Ciocalteu reagent (2 N) diluted to 1N, Deionized water.

• Proteins: BSA (standard) and sample.

• Glasswares: Beakers, cuvette (specify), pipettes (10ml), measuring cylinder (100ml), test tubes, glass stirrer.

Procedure

Standard Bicinchoninic Acid Preparation

• Reagent A Components: An aqueous solution of 1% BCA-Na_2, 2% $Na_2CO_3.H_2O$, 0.16% Sodium-tartrate, 0.4% NaOH, and 0.95% $NaHCO_3$. If needed, appropriate addition of NaOH (50%) or solid $NaHCO_3$ is made to reagent A to adjust the pH to 11.25.

• Reagent B Components: 4% $CuSO_4.5H_2O$ in deionized water.

• Standard Working Reagent (S-WR) (prepared weekly or as needed): Mix 100 vol of Reagent A with 2 vol of Reagent B. S-WR is apple green in colour.

Standard Bicinchoninic Acid Preparation (For micro-assay)

• Micro-Reagent A (MA) components: An aqueous solution of 8% $Na_2CO_3.H_2O$, 1.6% NaOH, 1.6% Sodium tartrate, and sufficient $NaHCO_3$ to adjust the pH to 11.25.

• Micro-Reagent B (MB) components:

4% BCA-Na_2 in deionized water.

• Micro-Reagent C (MC) components: 4 vol of 4% (aq) $CuSO_4.5H_2O$ plus 100 vol of Micro-Reagent B.

• Micro-Working Reagent (M-WR) consists of 1 vol of MC plus 1 vol of MA. MA and MB are stable indefinitely at room temperature, but MC and M-WR should be prepared as needed.

MacroAssay

1. Pass all the prepared solutions through a 1-pm filter to remove insoluble debris associated with the salt.

2. Mix 1 vol of the sample (standard or unknown) with 20 vol of S-WR in a test tube.

3. Colour development proceeds immediately, even at room temperature, but it can be greatly accelerated by incubating the tubes in a constant temperature water bath. In this respect, the temperature chosen for the colour development is directly related to the desired sensitivity.

4. The incubation protocols can be varied as follows: (i) room temperature for 2 h (ii) 37°C for 30 minutes (iii) 60°C for 30 minutes.

5. After choosing the incubation temperature, cool the samples to room temperature and measure their absorbances at 562 nm versus a reagent blank using a spectrophotometer.

6. The concentration of unknowns can be then determined from a plot of concentration vs absorbances obtained for the standard protein solutions.

MicroAssay (0.5-10 pg/ml)

1. To run the assay, mix 1 volume of sample or standard with 1 volume of M-WR in a test tube and incubate the tubes for 60 min at 6O°C.

2. Cool to room temperature, and measure the absorbance vs the blank at 562 nm.

3. Determine the concentration of unknowns from a standard curve.

Results

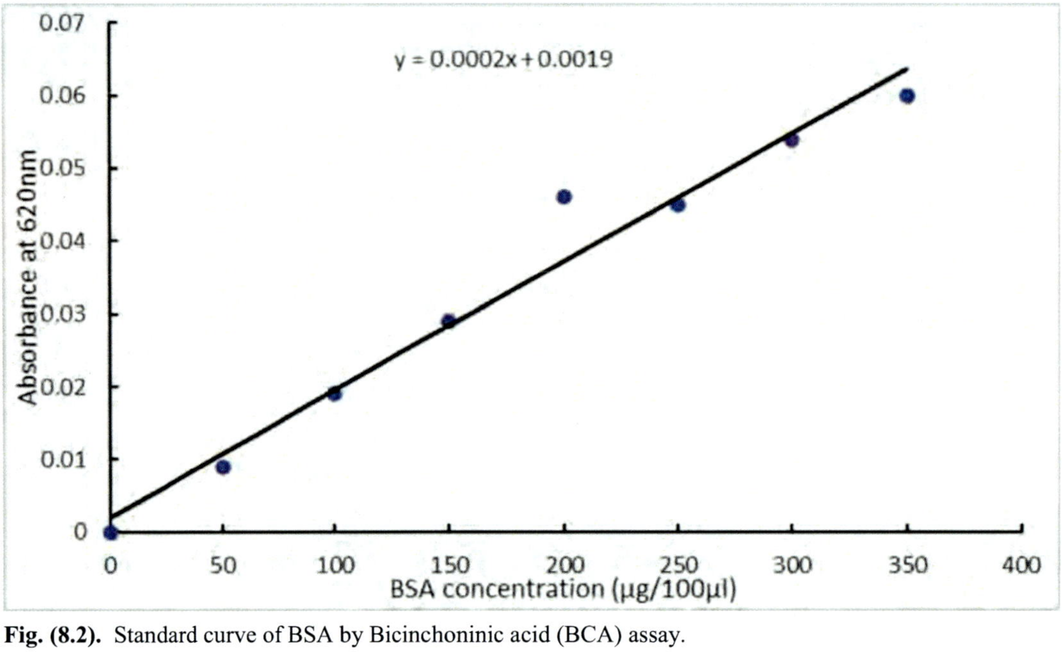

Fig. (8.2). Standard curve of BSA by Bicinchoninic acid (BCA) assay.

Calculation

Refer to Eqn. 3 from Chapter 3 for protein concentration computation by Biuret assay.

Results Interpretation

• By plotting an unknown sample's absorbance on the standard curve and relating that absorbance to the standard, the concentration of the solution can be calculated.

• Each dilution is averaged and plotted on a graph, and a trendline is added.

• Though the points may have some degree of deviation, they should be generally linear with a slight downward curve.

• Unknown concentrations are determined by relating their absorbances to the absorbances on the graph.

• Only use the unknown dilutions where the absorbance falls between the bounds of this graph (125-2,000 µg/ml or 5-250 µg/ml, depending on which set of standard dilutions was used).

• Plot the absorbance of each unknown dilution on the standard curve (Fig. **8.2**).

Troubleshooting

• If the substrate seems incompatible, try diluting/dialyzing the sample using the same buffer used for standards if still, the results are not as expected; make use of the Bradford assay.

• If no colour is visualized even if a sample is present, indicating that there is the presence of metal ions that acts as metal chelators. The solution to this problem would be dialyzing the solution using membranes or preparing a fresh reagent with reduced copper sulfate solutions.

• If the sample tube appears darker than usual, there is a presence of a reducing agent possessing thiol groups or catecholamines or lipoproteins; this can be resolved by dialyzing the sample or preferably using a fresh sample that is free from interference. Sometimes the dark colour of the sample is also due to high concentrations of protein in the sample. It is suggested to dilute the samples.

• There is a possibility that the colour developed in the sample is lighter as compared to the standard ones; this is mainly because of the presence of interference of an acidic or alkaline buffer which can be resolved by dialyzing the sample or preparing a fresh sample.

Did you know?

- The bicinchoninic acid (BCA) assay was invented in 1985 by Paul K. Smith at Pierce Chemical Company, now part of Thermo Fisher Scientific is a biochemical assay for determining the total concentration of protein in a solution.

BIURET METHOD

Aim

Quantitative estimation of proteins by the Biuret method.

Principle

The addition of a cupric salt to an alkaline solution of a protein produces a reddish-violet colour. This reaction is known as the biuret reaction. The general shape of the absorption spectrum of the biuret-protein complex is the same for all serum proteins, showing strong absorption in the ultraviolet and yellow-green, but the exact wavelengths of the absorption maxima and minima and the specific absorptivity depend upon the composition of the biuret reagent, nature of the protein and its concentration (Fig. **8.3**) [2]. The absorptive maximum is at 540 nm for serum proteins using the biuret reagents.

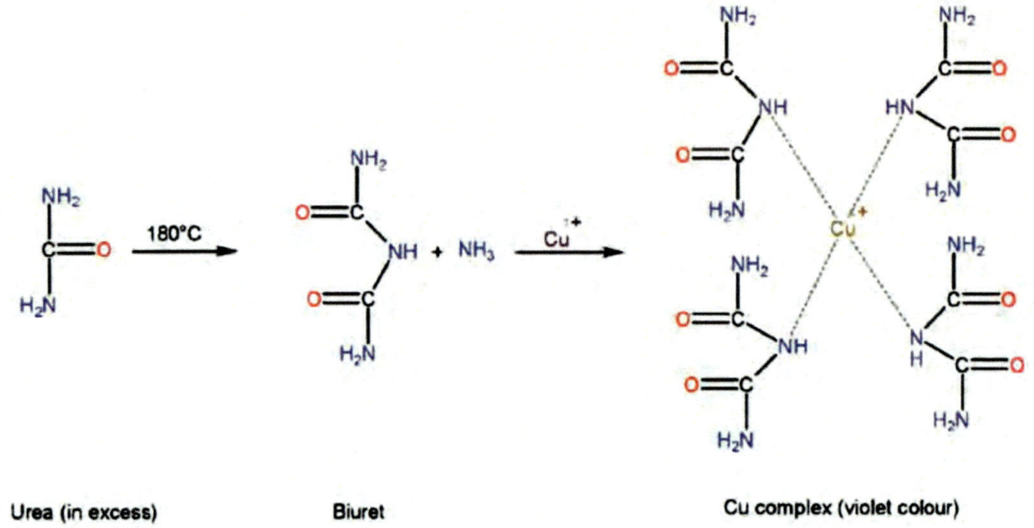

Fig. (8.3). Biuret Reaction.

Reaction range: 5-150 mg/mL.

Requirement

1. Reagents: Biuret Reagent.

2. Protein Standard: 5 mg BSA/ml.

3. Apparatus and Glasswares required: Test tubes, Pipettes, Cuvettes.

Procedure

Preparation of Biuret Reagent

• Dissolve 3 g of copper sulphate ($CuSO_4.5H_2O$) and 9 g of sodium potassium tartrate in 500 ml of 0.2 mol/litre sodium hydroxide; add 5 g of potassium iodide and make up to 1 litre with 0.2 mol/litre sodium hydroxide.

1. Pipette out 0.0, 0.2, 0.4, 0.6, 0.8 and 1 ml of working standard into the series of labelled test tubes.

2. Pipette out 1 ml of the given sample in another test tube.

3. Make up the volume to 1 ml in all the test tubes. A tube with 1 ml of distilled water serves as the blank.

4. Now add 3 ml of Biuret reagent to all the test tubes including the test tubes labelled 'blank' and 'unknown'.

5. Mix the contents of the tubes by vortexing/shaking the tubes and warm them at 37 °C for 10 min.

6. Now cool the contents to room temperature and record the absorbance at 540 nm against blank.

7. Then plot the standard curve by taking a concentration of protein along X-axis and absorbance at 540 nm along Y-axis.

8. Then from this standard curve calculate the concentration of protein in the given sample (Fig. **8.4**).

Results

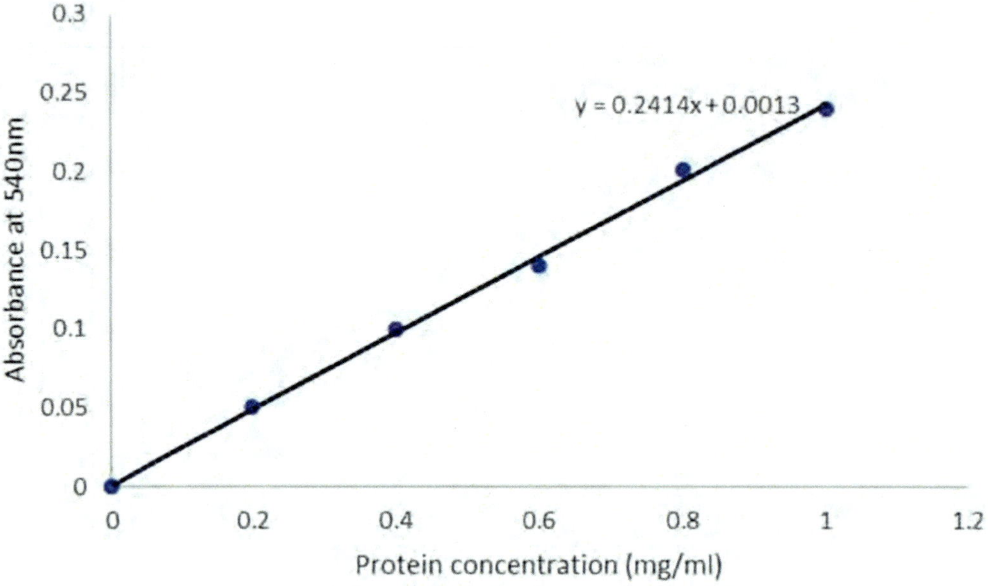

Fig. (8.4). Standard curve for the determination of protein concentration by the Biuret method.

Calculation

Refer to Eqn. 3 from Chapter 3 for protein concentration computation by Biuret assay.

The method has adequate sensitivity for protein concentrations found in normal and pathological sera. Below about 1 g per dl, the precision worsens as the absorbance of the reagent forms a large part of the total absorbance. The addition of a larger volume of a dilute protein solution (*e.g.*, cerebrospinal fluid) changes the concentrations of the constituents of the biuret reagent. The reagent should, therefore, be reformulated and the optimum composition determined, according to the range of protein concentrations to be measured. A reagent designed for blood serum protein concentrations may not necessarily be suitable for other fluids.

Advantages

- Biuret Assay for the quantification of proteins is widely used for the determination of the concentration of proteins in serum, mainly because of its convenience and good precision.

• The biuret reaction for proteins provides a simple and precise method for measuring serum proteins.

• Several stable biuret reagents are available.

Disadvantages

• Haemoglobin is the only important cause of interference that cannot be minimized by the use of a sample blank in the biuret assay.

Troubleshooting

• Few substances in blood serum interfere with the biuret reaction. Marked turbidity owing to lipemia may cause erroneously high results, while the use of a serum blank may overcorrect, leading to low results.

• Errors owing to lipemia can be reduced (1) by adding urea to the biuret reagent to a final concentration of 6 mol per 1 (2) by adding potassium cyanide solution to decolourize the biuret reagent (3) by extraction of the lipid from the serum using ether.

Did you know?

• The biuret test, also known as Piotrowski's test, is a biochemical test for detecting the presence of peptide bonds. The biuret reaction can be used to assess the concentration of proteins because peptide bonds occur with the same frequency per amino acid in the peptide.

• The intensity of the colour, and hence the absorption at 540 nm, is directly proportional to the protein concentration, according to the Beer-Lambert law.

• Despite its name, the reagent does not contain biuret ($(H_2N\text{-}CO\text{-})_2NH$). The test is named so because it also gives a positive reaction to the peptide-like bonds in the biuret molecule.

BRADFORD ASSAY

Aim

Quantitation of protein using Bradford assay.

Principle

Bradford protein quantitation assay works based on dye-binding, *i.e.*, Coomassie brilliant blue to proteins. Coomassie brilliant blue exists in two forms; the red colour is converted to blue upon binding to proteins. The quantity of protein can be estimated by determining the amount of dye in the blue ionic form (Fig. **8.5**). This is usually achieved by measuring the absorbance of the solution at 595 nm [4]. The protein-dye complex has a high extinction coefficient hence it leads to greater sensitivity in measuring protein.

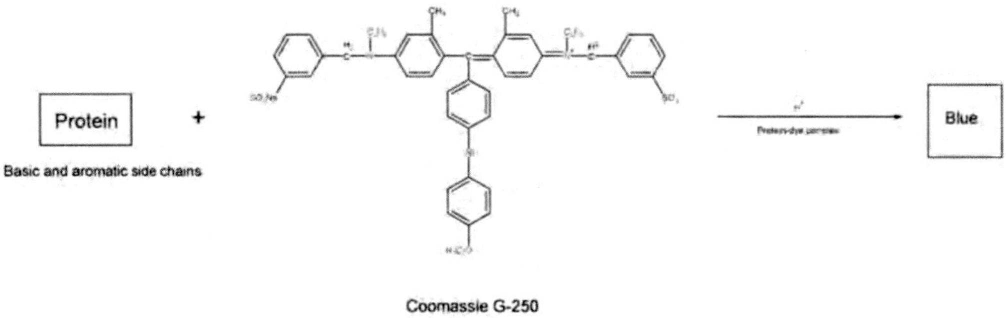

Fig. (8.5). Basic reaction under Bradford Assay.

Range of protein concentrations detected: 20-2000 μg/ml.

Requirements

Reagents

Coomassie Brilliant Blue G-250, 2-Mercaptoethanol, Triton X-100, Sodium Dodecyl Sulphate, Hemosol, Bovine Serum Albumin, Ethanol, 0.15 M Sodium Chloride, Phosphoric Acid.

Protein Solutions

Haemoglobin, Human Serum Albumin, Chymotrypsinogen A, Cytochrome c.

Procedure

Protein Preparation

- Prepare protein in 0.15 M NaCl.

- Concentrations were determined for bovine serum albumin, human serum albumin, chymotrypsinogen A, and cytochrome c spectrophotometrically in a Bausch and Lomb Spectronic 200 UV spectrophotometer based on ϵ280 = 6.6, 5.3, 20 and 17.1 respectively.

Preparation of Bradford Reagent

• Dissolve 100mg Coomassie Blue G-250 in 50 ml 95% ethanol.

• Add 100 ml 85% (w/v) phosphoric acid to that solution.

• Add distilled water to make up the volume to 1000 ml.

Protein Assay (Standard Method)

• Add 0.1 ml of prepared protein solution in test tubes.

• Add 5 ml of dye solution to the test tubes and mix it properly.

• Allow it to react for 2 minutes, and before 1 hour, check absorbance.

• Check the absorbance at 595 nm against a reagent blank prepared from 0.1 ml of the appropriate buffer and 5 ml of protein reagent.

• The weight of the protein was plotted against the corresponding absorbance resulting in a standard curve used to determine the protein in unknown samples.

• Plot the graph of the absorbance *vs.* weight of the protein.

MicroProtein Assay

• Add 0.1 ml protein solutions in test tubes and adjust the volumes of the samples by adding 0.1 buffer.

• Add 1 ml of dye solution and mix it by inversion or by the use of a vortex.

• Measure the absorbance at 595 nm against a reagent blank prepared from 0.1 ml of the appropriate buffer and 1 ml of protein reagent.

• Plot the graph of the absorbance *vs.* weight of the protein (Fig. **8.6**).

Results

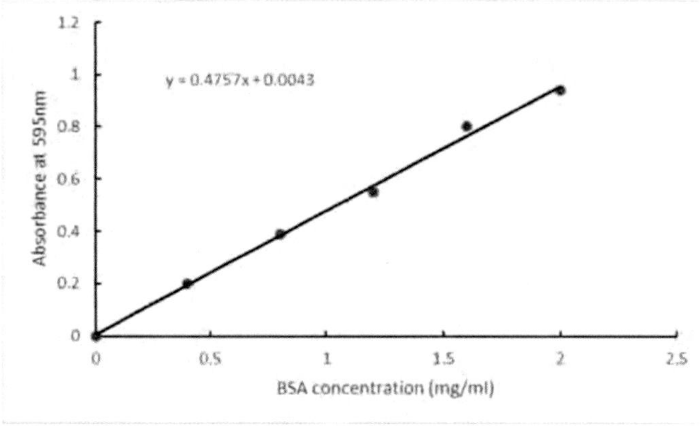

Fig. (8.6). Standard Curve of BSA by Bradford Assay.

Calculation

Refer to Eqn. 3 from Chapter 3 for protein concentration computation by Biuret assay.

Applications

1. To determine the residual protein content (RPC) in cell culture supernatants.

2. Only proteins with Tryptophan, Tyrosine and Phenylalanine are detected using this method due to the presence of aromatic rings.

Troubleshooting

• Precipitation Observed: The presence of detergents usually causes precipitation, which can be averted by diluting/dialyzing the protein sample.

• The absorbance of standards is lower than usual. This could be due to multiple reasons:

• The Bradford assay is either way past its expiry date or is not properly stored. The Bradford reagent must be stored for 12 months (maximum) at 4°C, once prepared.

• Check with the dilutions; if needed, perform those again.

• Check the temperature before using the Bradford reagent since a lower temperature reagent tends to result in absorbance at an incorrect wavelength.

• Samples appear darker than usual, indicating that pH is more towards the alkaline range. Dilute/dialyze the samples using a buffer.

Advantages

• Bradford is a rapid and sensitive assay to quantify proteins.

• The protein-dye-bound complex forms very rapidly and remains the same for a longer time. Bradford assay is a quick and sensitive technique making it the most attractive assay for protein quantitation.

• Protein quantification methods like Lowry are often subjected to interference due to compounds like ions, and carbohydrates; Methods like Biuret are subjected to interference due to ammonia, tris, and glycerol. Bradford assay solves all the issues mentioned earlier and is casicr to automatc.

Disadvantages

• The Coomassie Brilliant Blue G-250 binds more strongly to arginine and lysine residues of proteins and lesser to histidine, tryptophan, tyrosine, and phenylalanine; due to which there will be variations in result.

• Proteins with poor acid solubility cannot be assayed with this reagent.

Did you know?

• Bradford assay was originally described by Dr. Marion Bradford in 1976.

LOWRY METHOD

Aim

Quantitative estimation of protein by Lowry method.

Principle

Lowry assay uses a reagent named Lowry reagent. It comprises two solutions. One of which is a commercially available Folin-Ciocalteu reagent, and the other is prepared using a few chemicals, which include copper sulfate, sodium hydroxide, sodium potassium tartrate, and sodium carbonate. The principle behind this method is based on the reactivity of the peptide nitrogen bonds present with the copper [II] ions under alkaline conditions and subsequent reduction of the Folin Ciocalteu phosphomolybdic-phosphotungstic acid to hetero-poly molybdenum blue by the copper-catalyzed oxidation of aromatic acids (Fig. **8.7**) [3]. The pH of the reaction should be maintained between 10-10.5.

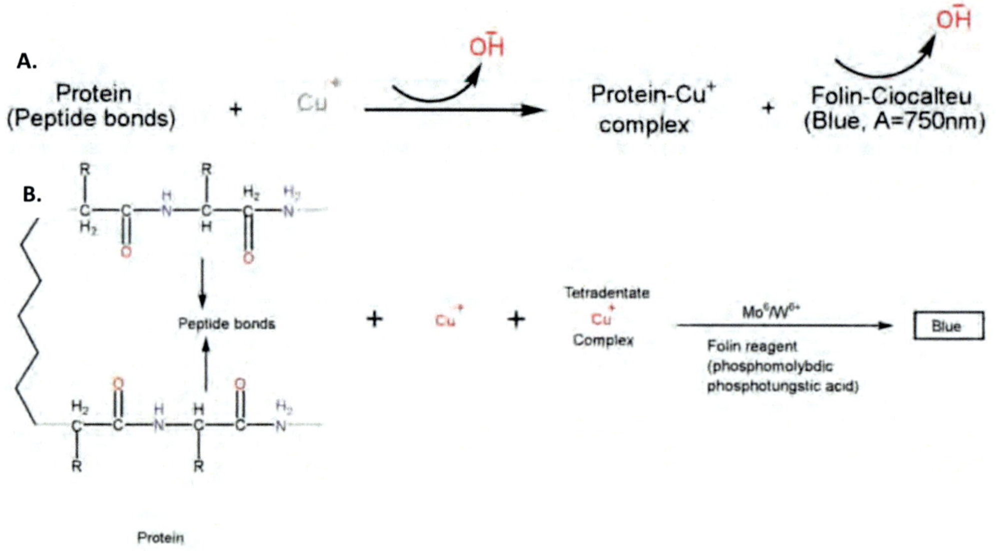

Fig. (8.7). A. - Basic reaction under Bradford Assay **B**. – Working of Bradford Assay.

Range: 10-1000 ug/ml.

Requirements

Reagent I: Lowry Reagent (or) Alkaline Copper Sulfate Solution

The following solutions are prepared:

SOLUTION A: 2% sodium carbonate in 0.1N NaOH.

SOLUTION B: 0.5% copper sulfate solution in 1% sodium potassium tartrate solution (to be prepared fresh).

Mix 50 ml of solution A with 1 ml of solution B, just before use.

Reagent II: Folin-Ciocalteau Reagent

This is commercially available and must be diluted with an equal volume of water just before use.

Standard Protein Solution

Dissolve 10mg of BSA (as it is easily available, cheap, and improved purity) in 100 ml of distilled water in a volumetric flask. (For concentration-100ug/ml).

Procedure

1. Take various concentrations of standard BSA working standard in 5 test tubes (Volumes from 0ml, 0.4ml, 0.8ml, 1.2ml, 1.6ml, and 2.0ml) and makeup to 2ml using distilled water.

2. The test tube with 2 ml distilled water serves as a blank.

3. Add 4.5 ml of Reagent I and incubate for 10 minutes.

4. After incubation, add 0.5 ml of Reagent II and incubate for 30 minutes.

5. Measure the absorbance at 660 nm and plot the standard graph.

6. Estimate the amount of protein present in the given sample from the standard graph (Fig. **8.8**).

Results

It is 10 or 20 times more sensitive than the measurement of the ultraviolet absorption at X = 280 rnp and is much more specific and much less liable to disturbance by turbidites. This assay is somewhat simpler, as well as much easier to adapt for small-scale analyses.

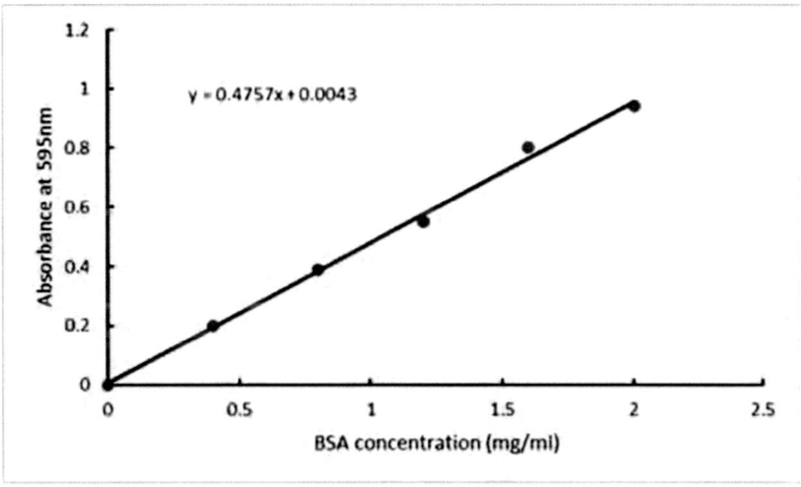

Fig. (8.8). Standard curve used for the Lowry protein assay.

Calculation

Refer to Eqn. 3 from Chapter 3 for protein concentration computation by Biuret assay.

Troubleshooting

• Samples appear darker than usual, indicating that pH is more towards the alkaline range. Dilute/dialyze the samples.

Applications

The reasonable applications of the Copper-Folin reaction include:

(1) measurement of protein during enzyme fractionations, *etc.*,

(2) measurement of mixed tissue proteins, particularly when absolute values are not needed,

(3) measurement of very small absolute amounts of protein, or highly diluted protein (*e.g.*, spinal fluid) or protein mixed with coloured substances or other nitrogen-containing substances and,

(4) analyses of large numbers of similar protein samples, such as antigen-antibody precipitates.

Did you know?

• Lowry assay was proposed by Oliver H. Lowry in 1951. His 1951 paper describing the technique is the most highly cited ever in the scientific literature, cited over 300,000 times.

• The total protein concentration is exhibited by a colour change of the sample solution in proportion to protein concentration, which can then be measured using colourimetric techniques.

ADVANCED TECHNIQUES FOR PROTEIN QUANTIFICATION

Fluorescence

To perform advanced fluorescent techniques for the quantitation of proteins, 3 major fluorophores used for the estimation of proteins by advanced fluorescence techniques have been described below: Epicocconone and Nanoorange.

Fluoroprofile

Principle

Fluoroprofile is a convenient and sensitive method for measuring protein concentration in solution, especially in 2-D gel sample buffers. It uses epicocconone to measure protein concentration in the solution. High-throughput automation can be performed with this assay, and the protein can be further analyzed biologically or chemically.

On binding with proteins under assay conditions, epicocconone yields an intense orange fluorescence suitable for quantitative measurement using an excitation range of 300– 550 nm (local excitation maxima are at 390 and 520 nm), and an emission maximum at 605 nm (Fig. **8.9**) [5].

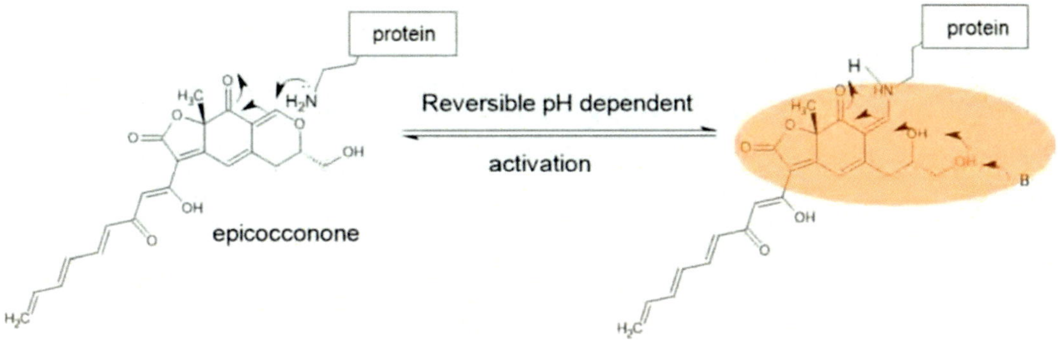

Fig. (8.9). Orange fluorescence emitted on the binding of proteins with epicocconone.

Range of protein estimation: 0.04 to 200 µg/mL.

Advantages

1. The measured light is of a different wavelength than the incident light, so fluorescence-based assays provide higher sensitivity, lower background, and greater dynamic range than colourimetric assays.

2. It offers a wide linear range in combination with high sensitivity.

3. The fluoro profile protein quantitation is suited for use in a wide range of instruments, like the Molecular Dynamics Typhoon scanner, the BMG Fluostar, and the Perkin Elmer LS50B.

4. This assay is quite robust to environmental conditions as incubation of labelled protein in microtiter plates, even at 30°C, had little effect on fluorescence intensity and linearity.

Disadvantages

1. A concentration of > 10 mg/mL of native iron-rich proteins such as myoglobin, cytochrome C, and holo-transferrin will not allow Fluoroprofile to maintain its linearity.

2. Proteins with very low pI, such as amyloglucosidase (pI 4.17), have slightly reduced fluorescent intensities.

NanoOrange Fluorescence Protein Estimation

This fluorescence method is used to quantify total protein using **NanoOrange** Protein Quantitation.

Principle

NanoOrange is based on a hydrophobic **cyanine dye** where the emission intensity is proportional to the amount of SDS–protein complex in the solution. In this assay, the binding of the compound NanoOrange with protein results in the formation of a fluorescent moiety, that does not lead to aggregation or loss of signal over time [5].

Range of protein estimated: 10ng to 10 μg/mL.

Advantages

1. The reaction is sensitive to the nanogram level, with fluorescent intensity values at very low concentrations.

2. The samples can be read hours later with no loss of sensitivity if protected from light due to the SDS detergent in the working solution maintaining the denatured condition of the protein following heating.

Disadvantages

1. Cyanine dyes aggregate easily, even in dilute solutions, resulting in calibration data that are sigmoidal, making quantitation at low or high concentrations prone to error.

2. NanoOrange is sensitive to some contaminants commonly found in protein preparations such as DNA, detergents, and some salts (*e.g.*, NaN_3, Na_3PO_4, $(NH_4)_2SO_4$, NaOH, HCl, and EDTA).

Accu Orange Fluorescence Protein Estimation

Principle

This fluorescence method is used to quantify total protein using **AccuOrange** Protein Quantitation. AccuOrange is recommended for quantifying purified protein or antibody samples. The fluorogenic dye binds primarily through hydrophobic interactions with proteins. Heat denatures proteins, and dye binds to their hydrophobic pockets after cooling. Fluorescence can be detected at 480 or 598 nm [5].

Range of Protein Estimation

0.1-15 ug/mL

Advantages

1. The assay is more sensitive than traditional protein quantitation assays.

2. It shows superior linearity and reproducibility than the NanoOrange protein quantitation assay.

3. The assay also shows minimal variability between different proteins and has a stable fluorescence signal for up to 16 hours.

Disadvantage

1. The AccuOrange™ assay has a low tolerance for non-ionic detergents and is not recommended for use with cell lysates containing Triton X-100, sodium deoxycholate, CHAPs, or other non-ionic detergents.

CONCLUSION

In this chapter, some basic standard methods and advanced techniques for protein quantification have been discussed. These basic techniques are essential and can be performed routinely in the lab. However, the advanced techniques are more accurate but expensive to perform.

VIVA VOCE

1. What should be used as a blank in BCA assay?

Ans: In this case, Reagent A and Reagent B. Do not add the sample in the blank tube.

2. I have to isolate RNA from an insect sample. How can I use the BCA assay in this protocol?

Ans: Once the RNA has been isolated, a BCA assay can be performed to check for protein contamination. If protein is detected, then the sample can be centrifuged again to get rid of proteins.

3. What is the importance of BSA in this assay? What could be the substitute protein for BSA?

Ans: Bovine Serum Albumin is used as a standard protein sample; a range of concentrations of BSA is prepared to quantify the amount of protein present in our sample. BSA is particularly attractive because of its low cost, high purity and ready availability.

Egg albumin is very similar in amino acid content to bovine serum albumin (BSA) and can be an excellent substitute for BSA. The tyrosine and tryptophan in BSA are 5.1 g/100g and 0.6 g/100g of protein, while in Egg albumin, are 3.9 g/100 g and 1.2 g/100 g of protein, respectively, hence the concentration range may be similar to each other.

4. What is the colour change observed in this reaction?

Ans: Blue to purple due to the presence of proteins.

5. What is the reaction associated with the Biuret assay?

Ans: When a protein reacts with copper(II) sulfate (blue), the positive test is the formation of a violet-coloured complex.

6. Which is the dye used in the Bradford assay? What is the colour change observed.

Ans: Coomassie Brilliant Blue is the dye. The colour changes from red to blue.

7. Describe the principle of the Bradford Assay?

Ans: This method relies on the dye binding to protein; the more intense the colour the more the concentration of protein in that sample.

8. Which other protein except BSA can be used as a standard in Bradford Assay?

Ans: IgG, if a direct comparison of two different protein assays is being performed, the same standard should be used for both procedures.

9. I had run a protein sample obtained after purification from E. coli and had run the sample on SDS-PAGE. There were no bands obtained. How can I use Lowry assay in this situation?

Ans: Elute the sample from the gel and check for protein concentration using the Lowry assay.

10. I have isolated a protein from fungal species, which is known to have a lot of polysaccharide contamination. How am I supposed to determine the amount of protein I have extracted?

Ans: The protein content can be quantified using the Lowry assay. If it is shown less than the expected concentration then, carbohydrate quantification tests can be used to check for polysaccharide contamination.

11. How does Protein-to-Protein variation affect the specificity of the FluoroProfile protein quantification assay?

Ans. A variety of proteins were tested using the assay at a range of concentrations, and minimal variation in reaction response was observed. Measuring proteins with heme was difficult only in a few cases. The determination of iron-rich proteins is only accurate up to a concentration of 10 g/ml.

12. What is meant by the pI of protein?

Ans. The isoelectric point (pI) of a solution is the pH at which the net charge of a protein becomes zero. At a solution pH that is above the pI, the surface of the protein is predominantly negatively charged, and therefore like-charged molecules will exhibit repulsive forces. Similarly, at a solution pH below the pI, the protein surface is predominantly positively charged, and repulsion between proteins occurs. However, positive and negative charges are balanced at the pI, reducing electrostatic repulsive forces respectively, and attraction forces operate more strongly, causing aggregation and precipitation. The pI of most proteins is in the pH range of 4 to 7.

REFERENCES

[1] Smith, P.K.; Krohn, R.I.; Hermanson, G.T.; Mallia, A.K.; Gartner, F.H.; Provenzano, M.D.; Fujimoto, E.K.; Goeke, N.M.; Olson, B.J.; Klenk, D.C. Measurement of protein using bicinchoninic acid. *Anal. Biochem.,* **1985**, *150*(1), 76-85. [http://dx.doi.org/10.1016/0003-2697(85)90442-7] [PMID: 3843705]

[2] Kumar, V.; Gill, K.D. *To determine the quantity of proteins in urine sample using biuret reaction. InBasic concepts in clinical biochemistry: A practical guide*; Springer: Singapore, **2018**, pp. 39-41. [http://dx.doi.org/10.1007/978-981-10-8186-6_9]

[3] Lowry, O.; Rosebrough, N.; Farr, A.L.; Randall, R. Protein measurement with the Folin phenol reagent. *J. Biol. Chem.,* **1951**, *193*(1), 265-275. [http://dx.doi.org/10.1016/S0021-9258(19)52451-6] [PMID: 14907713]

[4] Knight, M.I.; Chambers, P.J. Problems associated with determining protein concentration: A comparison of techniques for protein estimations. *Mol. Biotechnol.,* **2003**, *23*(1), 19-28. [http://dx.doi.org/10.1385/MB:23:1:19] [PMID: 12611266]

[5] Krohn, R.I. The colorimetric detection and quantitation of total protein. *Curr. Protoc. Toxicol.,* **2005**, *23*(1), 1-28. [http://dx.doi.org/10.1002/0471140856.txa03is23] [PMID: 20976675]

CHAPTER 9

Protein Purification

Sinchan Hait[1], **Sai Joshi**[2,*] and **Ankita Nitin Padhye**[3]

[1] *Department of Chemical Engineering, BITS-Pilani K.K. Birla Goa Campus, Goa-403726, India*

[2] *CSIR-NCL, Dr. Homi Bhabha Road, Pune - 411 008, India*

[3] *Food Engineering and Technology Department, Institute of Chemical Technology, Mumbai, Maharashtra, India*

Abstract: Protein purification is an essential step after protein isolation. Protein purification ensures the removal of all contaminants from the sample pool. Purification depends on the chemical and physical interactions of protein molecules and chemical moieties. These techniques are performed with tags and without tags. Tag purification involves ammonium sulfate precipitation and dialysis. Purification of proteins with tags utilizes affinity and size exclusion chromatography. There are various methods available for this, however, we will discuss some most common methods for protein purification.

Keywords: Ammonium sulfate precipitation, Affinity chromatography, Dialysis, Protein, Protein purification, Size exclusion chromatography.

INTRODUCTION

Protein purification techniques improve the protein quality in the sample by removing contaminants from it. Major techniques are performed with and without a tag for purification. Both techniques depend upon the physio-chemical interactions between the protein molecule and the chemical compound present in it. The techniques are discussed in detail below:

PURIFICATION OF DIFFERENT TYPES OF PROTEINS (WITHOUT TAG)

Ammonium Sulfate Precipitation

Aim

To use the ammonium sulfate precipitation method for protein purification.

* **Corresponding author Sai Joshi:** CSIR-NCL, Dr. Homi Bhabha Road, Pune - 411 008, India; E-mail: saijoshi1997@gmail.com

Pamela Jha (Ed.)

Principle

The ammonium sulfate precipitation method, also known as the salting out method, is a commonly used method for protein purification from a solution.

Most of the enzymes or proteins exist in cell fluids as soluble proteins; this solubility is attributed to the combined electrostatic effects of polar interactions of the protein with aqueous solvent and electrostatic forces between protein molecules and small aggregates of like charges [1, 3].

When the concentration of the ammonium sulfate increases, the small, highly charged ions of the salt compete with the proteins to bind to the water molecule. This removes the water molecule interactions or the 'hydration shell' around the protein and exposes the hydrophobic core of the protein to interact intermolecularly, thereby decreasing the protein solubility and resulting in precipitation [1, 3]. Hence precipitation of protein by the addition of salts like ammonium sulfate is known as 'salting out' (Fig. **9.1**).

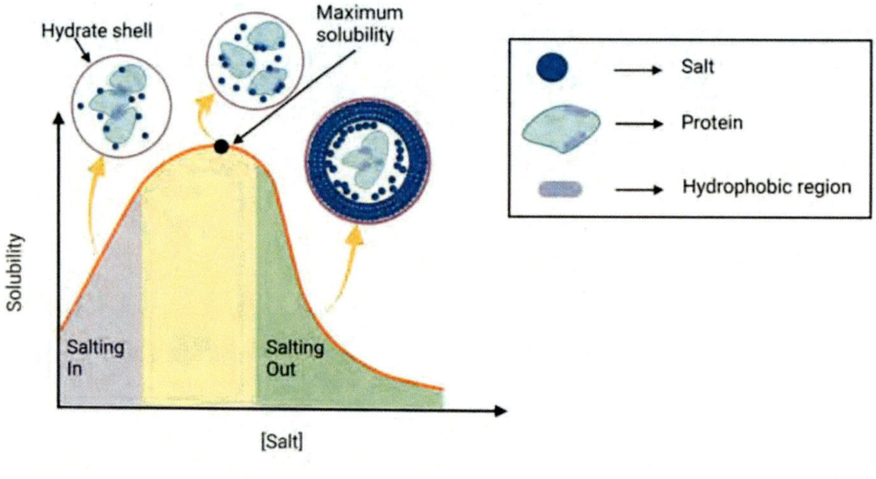

Fig. (9.1). Salting in and out curve.

Requirements

❖ ***Chemicals and Sample***

- Protein sample
- Ammonium sulfate
- Phosphate buffer saline

❖ ***Instruments***

- Centrifuge
- Refrigerator
- Weighing balance
- Magnetic stirrer (optional)

❖ ***Miscellaneous***

- Conical flasks
- Stirrer
- Petri dish.

Procedure

1. The ammonium sulfate for precipitation needs to be in uniform, dry form; for that, place the salt in a dish and dry heat at 120 ºC in an oven.

2. Centrifuge the solution at 10000 g at 4 ºC for 10 mins to get rid of any visible precipitate.

3. Collect the precipitated protein by centrifugation for 30 minutes at ≥ 5,000 g at 4°C. Decant the supernatant. Dissolve the pellet (generally in 10 times lower volume than the initial volume) in the appropriate buffer.

4. Prepare ammonium sulfate of 45% saturation, *i.e.*, 277 g ammonium sulfate/litre solution. Stir gently for a minimum of 4 hr and a maximum of 16 hr at 4 ºC.

5. Add the prepared ammonium sulfate slowly while stirring in the prepared protein solution very slowly. This will ensure that the local concentration around the site of addition does not exceed the desired salt concentration.

6. Once the ammonium sulphate is added, move the container to the refrigerator at 4 ºC for 6 hours for sufficient precipitation to occur.

7. Transfer to a conical tube and centrifuge the precipitate at 3000g for 30 minutes.

8. Carefully remove and discard the supernatant. Invert the conical tube and drain well.

9. Resuspend the pellet in PBS, pH 7.2. The solution will remain turbid, but discrete flocculent is not visible.

10. Remove residual precipitate and collect the clarified proteins by performing centrifugation with the following parameters: 25,000g at 4ºC for 10 mins.

11. Remove the clarified proteins to fresh tubes and store them at 4 ºC for up to one week.

Results

Half Saturation

The presence of globulin in the solution is indicated by the presence or formation of a white precipitate in the solution. It also indicates the success of the reaction. The absence of the white precipitate represents a negative result, indicating the absence of globulin in the sample.

Full Saturation

• If a white precipitate is formed in the solution under full saturation conditions, then it indicates that there is a presence of albumin in the sample. On the other hand, the absence of white precipitate is an absence of albumin in the sample.

Precautions

1. The ammonium sulfate for precipitation needs to be in uniform, dry form; for that place the salt in a dish and dry heat at 120 ºC in an oven.

2. Add the prepared ammonium sulfate slowly while stirring in the prepared protein solution very slowly. This will ensure that the local concentration around the site of addition does not exceed the desired salt concentration.

3. With solid ammonium sulfate, a mortar and pestle can be used to break up any lumps.

4. Use analytical grade as lower grade material is often contaminated with heavy metals.

5. The addition of ammonium sulfate acidifies the solution, so use at least a 50 mM HEPES or Tris buffer *etc.*, 5 mM EDTA can also be included.

6. Add solid ammonium sulfate slowly with gentle stirring; allow it to dissolve before adding more solid, and try to prevent foaming.

Advantages

1. Ammonium sulphate salts contain polyvalent ions, *i.e.*, sulphate ions which have a higher capacity to enhance the salting-out process.

2. The salting-out process using ammonium sulfate is not temperature dependent and hence wouldn't affect the protein in any way.

3. The preparation of 2-3M ammonium sulfate is stable for many years and can be stored since high concentrations of salt prevent proteolysis and bacterial action. Due to this, the process becomes cheaper.

4. The addition of salts is always preferable to minimize the final protein concentration.

Limitations

1. While using solid ammonium sulfate, lumps might be formed in the solution. To avoid that, mortar and pestle should be used.

2. The use of low-grade mineral salts might cause contamination with heavy metals that inhibit the precipitation process.

3. Online calculators can be used to determine the amounts of solid ammonium sulfate required for saturation.

4. The use of ammonium sulfate might cause acidification of the solution, which slows down the process; thus, Tris buffer can be added.

Protein Purification by Dialysis

Aim

To purify protein using dialysis.

Principle

Conventional dialysis separates small molecules from large molecules by allowing diffusion of only the small molecules through selectively permeable membranes. Dialysis is usually used to change the salt (small-molecule) composition of a macromolecule-containing solution or for the purification of any macromolecule by separating salt solution [1, 3].

Dialysis involves the separation of solutes by diffusion across the membrane from one liquid phase to another liquid phase based on molecular size and molecular conformation. A semipermeable membrane is used to contain the target protein. The target protein in the solution is placed into the dialysis tubing, and the dialysis tubing is sealed and placed into a large container of water. The salt in the dialysis tubing moves from high concentration to low concentration *via* passive diffusion. The final salt concentration in the protein solution is the weighted average of the salt concentration based on the volumes of the starting protein solution and the water. As the volume of the solution in the dialysis tubing decreases, the protein concentration increases [1, 3].

Factors Affecting Dialysis

The time required depends on factors that control the rate of diffusion of molecules.

Temperature

Since increasing the temperature speeds up the rate of diffusion, the procedure will proceed faster at room temperature than at 4°C. However, you will need to take the thermal stability of the molecule of interest into account when determining the optimal temperature.

Concentration and Molecular Weight of the Molecule

The higher the concentration, the faster the rate of diffusion. However, as the molecular weight increases, the rate of movement slows down considerably.

Surface Area and Thickness of the Membrane

The diffusion rate is directly proportional to the surface area of the membrane and is inversely proportional to its thickness.

Requirements

Sample

Protein solution.

Dialysis Buffer

NaCl, Tris-HCl, Glycerol, ddH2O.

Dialysis Bag

Dialysis bag with a few clips.

Procedure

Dialysis Buffer Preparation

• Prepare 3L of dialysis buffer (Dialysis can be divided into 3 stages, and each stage would need 1L of dialysis buffer.)

• Target concentration: NaCl-150 mM, Tris-HCl-20mM, glycerol 20% store at 4°C.

Dialysis Tube Preparation

• Cut the proper length of the dialysis bag. Soak it in the dialysis buffer.

Dialysis Process

• Load the sample into dialysis tubing, cassette or device and dialyze for 2 hours. You can perform this step at room temperature or 4°C.

• Clip the bottom of the dialysis bag with a metal clip to seal it well. Add the protein into the dialysis bag, and get rid of all bubbles. Clip the other end with a plastic clip and tie an Eppendorf onto it.

• Put a stir bar in the dialysis buffer and stir at 4°C.

• Change the dialysis buffer every 3 hours. The last stage of dialysis can be stored overnight.

• Collect samples from the dialysis bag.

• Centrifuge 13000 rpm, 4°C, 20 mins after centrifuging, obtain the supernatant and store at -20°C.

Precautions

• The dialysis bag should be suspended in the dialysis buffer; thus, the two sides of the dialysis bag would be clipped with different kinds of clips.

• The stir bar would attract the metal clip by magnetic force; the dialysis bag would rotate with the stir bar.

Troubleshooting

Problem

Protein precipitation is a common problem faced during protein purification using dialysis. The common causes and probable solutions is mentioned in Tables **9.1** and **9.2**.

Table 9.1. Common causes and solutions of protein precipitation.

Cause	Solution
• Low salt concentration in dialysis buffer. • High protein concentration is present in the sample. • Frequent and sudden pH changes. • The pH of the dialysis buffer may be close to that of the isoelectric point of the target protein.	• Use at least 350 mM concentration of salt in buffers. • Use at least 1000X excess dialysis buffer. • Perform the procedure at 4 degrees Celsius. Change the buffer every 3 hours. • If still not resolved, add 50% glycerol solution.

Table 9.2. Few problems and probable solutions of protein purification.

Problems	Solutions
The sample for protein purification is dilute.	Concentrate the solution using methods like Centrifugation, Lyophilisation and Precipitation.
The target protein from the sample is lost.	Use dialysis for purification only when the Molecular weight of the protein is 2X greater than the MWCO.
The purified protein contains salt impurities.	Reduce the volume of the protein to be purified at one time. Perform the desalting procedure using gel filtration chromatography.
The dialysis membrane does not float.	Make sure that the flask ring is attached properly and that the proper amount of solution is present in the flask.
Not all contaminants and impurities are removed.	Increase the dialysis time. Change the buffers frequently. Use a higher MWCO* membrane.

*MWCO is Molecular Weight Cut-off; defined as the molecular weight at which 90% of the macromolecular solute is rejected by the membrane. It is mostly used for characterizing dialysis or ultrafiltration membranes.

Did you know?

The first scientific descriptions of these procedures date back to the 19th century and came from the Scottish chemist Thomas Graham, known as the "Father of Dialysis." At first, osmosis and dialysis became popular as methods used in chemical laboratories that allowed the separation of dissolved substances or the removal of water from solutions through semipermeable membranes.

PURIFICATION OF DIFFERENT TYPES OF RECOMBINANT PROTEINS (WITH TAG)

Protein Purification by Affinity Chromatography

Aim

To purify the protein using Affinity Chromatography.

Principle

It relies on the specific and reversible binding of a protein to a matrix-bound ligand. The specific ligand of the desired protein or enzyme is immobilized on the insoluble matrix. This ligand reversibly binds to the protein of interest in the multi-component mobile phase. The protein of interest can be eluted by changing the composition of the column [1 - 4].

M + L <====> ML

(Macromolecule) (Ligand attached to the matrix)

All other compounds can therefore be washed away, and the compound can subsequently be recovered by displacement from the ligand.

Components in Affinity Chromatography

Matrix

An ideal matrix should be:

Exhibit proper and enough chemical groups which can covalently link the ligand and sustain the conditions of attachment.

Should be firm enough to sustain binding and subsequent elution of macromolecule.

React weakly or should not react at all with other molecules in the reactions. This will prevent non-specific adsorption.

Possess fine flow characteristics.

Ligand

The molecule to be purified will dictate the chemical nature of the ligand.

The ligand should possess a chemical group that will not be a part of the reversible binding of the ligand to the molecule of interest and also help in the attachment of the ligand to the matrix

The ligand should display a selectivity to closely related compounds having similar chemical nature.

Spacer Arm

A spacer arm is inserted between the ligand and matrix to forbid the interference between the attachment of the ligand to the matrix and the ability of the ligand to bind with the molecule of interest.

The spacer arm must be 6-10 carbon atoms long or equal. The length of the spacer arm is a crucial component in the success of the separation reaction.

Buffers for Elution

Buffer Elution

Buffer conditions are changed to reverse (weaken) the interaction between the target molecules and the ligand so that the target molecules can be eluted from the column.

pH Elution

The degree of ionization of charged groups on the ligand and or the bound protein is altered by a change in pH. This change in pH can affect binding sites and affinity by altering respective conformations. The limit of pH can be determined by the chemical stability of the matrix, ligand, and molecule of interest. A step decrease in pH is the most common way to elute bound substances.

Ionic Strength Elution

The exact mechanism for elution by changes in ionic strength will rely on the specific reaction between the ligand and molecule of interest.

Competitive Rlution

Selective eluents are often used to separate substances on a group-specific medium or when the binding affinity of the ligand/target protein interaction is relatively high. The eluting agent competes either for binding to the target protein or for binding to the ligand.

The Reduced Polarity of Eluent

Conditions are used to lower the polarity of the eluent and promote elution without inactivating the eluted substances. Dioxane (up to 10%) or ethylene glycol (up to 50%) is typical of this type of eluent.

Chaotropic Eluents

If other elution methods fail, deforming buffers, which alter the structure of proteins, can be used, *e.g.*, chaotropic agents such as guanidine hydrochloride or urea. Chaotropes should be avoided whenever possible since they are likely to denature the eluted protein.

Wash Buffer

Conditions that wash unbound substances from the column without eluting the target molecules or that re-equilibrate the column back to the starting conditions (in most cases the binding buffer is used as a wash buffer).

Affinity Tag Removal

Affinity tag enables the purification of any protein without any prior knowledge about its biochemical properties hence becoming the most essential tool for protein purification (Table **9.3**).

Table 9.3. Protein/Peptide Tags.

Protein to Purify	Ligand	Elute with
Antibody (Antigen-specific)	Antigenic Peptide	Free peptide
Antibody (Class-Specific)	Protein A or G or L or Protamine	Extreme values in pH
Polyhistidine-tagged protein	Nickel or Cobalt	Imidazole or Free Histidine
FLAG-tagged protein	FLAG-specific protein	FLAG peptide or Low pH
GST(Glutathione S-transferase)- tagged protein	Reduced Glutathione	Free Glutathione
Myc-tagged protein	Myc-specific antibody	Low pH
DNA-binding Protein	Heparin	High ionic Strength
Strep Tag II	Biotin	Biotin Analog

Functions of Protein Tags

Facilitate Recombinant Protein detection and hence purification.

Increase protein solubility and yield in some cases.

Promote appropriate protein folding.

Reasons for Tag Removal

To prevent crystallization of the desired recombinant protein.

To obtain pure protein solutions. The tags might interrupt the downstream processing of the protein.

In some cases, the protein conformation (shape and size) is severely affected by the presence of a tag.

Basic Steps Involved in the Removal of the Tag

1. Position a Protease specific sequence between the affinity tag and the target protein.

2. Process and purify the protein.

3. Treat the obtained protein solution with the specific protease.

4. Centrifuge and separate the untagged protein from the protease solution. The untagged protein can also be separated using electrophoresis Table **9.4**.

Table 9.4. Some common enzymes used for Tag removal.

Enzyme	Amino Acid Sequence	Attributes
Enterokinase	DDDD**K**	Secondary sites at basic amino acids require Ca2+ as a cofactor.
Aminopeptidase M	Exopeptidase	Cleaves at the N-terminal of protein; does not cleave X-P.
TEV - protease	EQLYFQ**G**	Commonly used for His-Tag, MBP-tag removal.
Thrombin	LVPR**GS**	Commonly used for Biotin-Tag removal.
- 3C Protease - PreScission	- ETLFQ**GP** - LEVLFQ**GP**	Contains GST tag for removal of the protease.
Sortase A	LPET**G**	Ca2+-induction of cleavage requires an additional affinity tag (*e.g.*, his-tag) for on-column tag removal.
Intein	Self-Cleavable	Artificial amino acids are left after cleavage in some applications.
TAGzyme or DAPase	Exopeptidase	Cleaves C-terminal; used for His-Tag removal.

*The amino acid sequences in bold stay even after cleavage.

Requirements

❖ ***Biological Requirements***

- Sample Supernatant

❖ ***Chemicals***

- PBS Buffer
- Tris base buffer
- Citric Acid

❖ ***Equipment***

- Magnetic Stirrer
- Refrigerated Stirrer
- Spectrophotometer
- Column
- Dialysis tubing and clamps

❖ ***Miscellaneous***

- Absorbent paper
- Forceps
- Glass pipette
- Pipette Pump
- Centrifuge tube
- 0.45μm filter
- Syringe
- Pasteur pipette
- Test tubes

Procedure

Phosphate Buffer Saline

Weigh 0.23 g NaH_2PO_4 (anhydrous; 1.9 mM),1.15 g NaH_2PO_4 (anhydrous; 8.1 mM), 9.00 g NaCl (154 mM). Add water to a volume of 900 ml. Adjust to the desired pH using 1 M NaOH or 1 M HCl. Add water to get a final volume of 1 liter.

HCl

Mix in the following order: 913.8 ml H_2O and 86.2 ml concentrated HCl.

NaOH (10 M)

Dissolve 400 g NaOH in 450 ml H_2O. Add H_2O to 1 liter.

1. Centrifuge the protein supernatant at 4°C.

2. Filter the protein supernatant using the filter (15 microns).

3. Prepare the affinity column by loading the support matrix and washing it with a PBS buffer.

4. Layer the column with the particular ligand along with the spacer arm onto a resin bed.

5. Elute the column with a suitable washing buffer.

6. Collect and dialyze (Check dialysis for the process) the eluent to obtain the protein of interest.

Precautions

- Check for protease inhibitors while tagging removal.

- Temperature, the concentration of protease, length of incubation with protease, and salt concentration should be optimized.

- There should be a significant difference between the Molecular weight of proteins and the tag making the process of untagged protein separation a lot easier.

- The resin for fractionation should be selected according to the size of the proteins to be separated.

• The sample volume should be less than 5% of the column volume to ensure elution as a sharp peak; the higher the ratio of sample to column volume, the lower the resolution. The sample buffer need not be the same as the chromatography buffer. The problems encountered in protein purification is summarized below (Table **9.5**).

Table 9.5. Some problems, their causes and probable solutions in protein purification.

Problem	Cause	Solution
Protein Does not bind	a. The tag was not translated or is not accessible b. Binding conditions are not appropriate c. Not enough time was allowed for binding d. The affinity between ligand and tag is very high	a. Check the plasmid sequence or move the tag to a different location b. Adjust buffer conditions c. The decreased flow rate or stop column allows incubation d. Increase concentrations of competitors (specificity) or stringency of conditions (non-specificity)
Protein does not elute	a. Protein aggregated on the column b. The flow rate is either too fast or too slow c. The column was not washed sufficiently	a. Adjust buffer conditions for more protein stability b. Adjust flow rate c. Wash with higher stringency buffer
Low-resolution	a. Protein aggregated on the column b. Elution conditions are not stringent enough c. Protein is unfolded or aggregated	a. Adjust buffer conditions for more protein stability b. Increase concentrations of competitors (specificity) or stringency of conditions (non-specificity) c. Adjust buffer conditions for more protein stability
Protein loses its activity during elution	a. A cofactor required for activity was removed during purification	a. Add cofactor in appropriate concentrations

Advantages

• High specificity.

• Target molecules can be obtained in a highly pure state.

• Single-step purification.

• The matrix can be reused rapidly.

• The matrix is solid and can be easily washed and dried.

• Give purified products with high yield.

• Affinity chromatography can also be used to remove specific contaminants, such as proteases.

Disadvantages

• Time-consuming method.

• More amounts of solvents are required, which may be expensive.

• Intense labour.

• Non-specific adsorption cannot be eliminated; it can only be minimized.

• Limited availability and high cost of immobilized ligands.

• Proteins get denatured if the required pH is not adjusted.

Protein Purification by Size Exclusion Chromatography

Aim

To purify proteins using Size Exclusion Chromatography.

Principle

Size exclusion chromatography is often known as gel filtration chromatography. It can be used to either fractionate molecules and complexes in a sample into fractions with a particular size range, to remove all molecules larger than a particular size from the sample, or a combination of both operations. Gel filtration chromatography can be used to separate compounds such as small molecules, proteins, protein complexes, polysaccharides, and nucleic acids when in an aqueous solution. When an organic solvent is used as the mobile phase, the process is instead referred to as gel permeation in this type of chromatography, and the molecules are separated based on their hydrodynamic radius, which is dependent on the size and shape of the protein [1, 3, 5].

The SEC relies on the ability of a given size of protein molecule to enter the uniformly sized pores of the stationary phase. The stationary phase and length of the column are used to determine the resolution of the proteins separated by the SEC.

The stationary phase consists of a porous matrix whose pore size is fixed according to the molecule to be separated. The beads or the matrix have a defined pore size range known as the fractionation range. The mobile phase containing the analytes to be separated is passed through the column in the form of a buffer. Molecules of the analytes that are greater than the pore size fail to enter the matrix and hence stay in the buffer. Since the pore size of the matrix is decided according to the desired protein molecule, the rest tend to remain in the buffer. The smallest molecule which remains in the buffer is known as the exclusion limit of the matrix [5].

Different stationary phases are commercially available, but the choice of stationary phase depends on the type of molecule to be separated. Following are some of the stationary phases (Table **9.6**).

Table 9.6. Stationary phases, their chemical composition and attributes.

Stationary Phase	Chemical Composition	Attributes
Sephadex	Cross-linked Dextran and Epichlorohydrin	1. Can be sterilized by autoclave 2. Quick buffer exchange and group separation are possible 3. Well suitable to determine the molecular weight 4. Specific types of Sephadex are available for use concerning solvents.
Sephacryl	Cross-linked allyl dextran and N, N methylene bisacrylamide	1. The recovery rate is very high 2. Suitable with organic as well as an aqueous solvent 3. Can separate molecules over a very broad range of molecular weights 4. Can be sterilized by autoclave
Sepharose	Cross-linked agarose	1. Autoclavable 2. High recovery rate
Superose	Highly cross-linked agarose	1. Compatible with viscous solvents 2. Suitable for hydrophobic interactions between protein and matrix 3. Works well with organic and aqueous solvents 4. Autoclavable
Superdex	Cross-linked agarose and dextran	1. High resolution and Recovery 2. Works with aqueous and organic solvents 3. Autoclavable

Application

- Buffer exchange and Desalting.
- Determination of the size of the molecule.

• Separation of bound from unbound radioisotopes.

• Fractionation of molecules and complexes within a predetermined size range.

• Removal of large proteins and complexes as well as small molecules like nucleotides, primers, dyes.

Types of Gel Filtration Chromatography

Group Separations	High-resolution Fractionation of Biomolecules
Group separation is employed when there is the removal of high or low molecular-weight compounds.	High-resolution fractionation can be used to isolate one or more components, separate monomers from aggregates, determine the molecular weight, or perform a molecular weight distribution analysis.

Apparatus

❖ Gel filtration medium

❖ Chromatography column

❖ Gel fractionation buffer

❖ Protein sample

Procedure

Buffer Preparation

Chromatography Buffer Stock Preparation

a. Dissolve 121.1g Tris in 600 mL water in a glass beaker.

b. Make the pH 7.5 with HCl once the tris is completely dissolved. Then make up the volume to 1000 mL using distilled water and store at 4°C.

Salt Stock Solution

a. Weigh 292.2 g NaCl and dissolve in 500 mL water.

b. Add water to a final volume of 1000 mL. Store at room temperature.

Chromatography Buffer

a. Incorporate 20 mL 1 M Tris, pH 7.5 with 20 mL 5 M NaCl and 960 mL water (see Note 4). Store at 4°C.

Gel Preparation

1. Add a dry volume of gel to a volume of gel fractionation buffer equal to twice the gel volume.

2. Stir the gel with glass and allow it to swell for 3 hours at room temperature.

3. Let the gel bed settle and discard the solution to remove any broken beads or cracks. Repeat the decantation 4-5 times maximum and then dilute to get the final slurry with 50% (v/v) settled gel and 50% (v/v) GF buffer.

Pack the Column

1. Ensure that the column is clean. Check the column for any cracks and damages.

2. Place the column on a laboratory stand vertically. Place an extension below the column to hold the excess gel slurry volume.

3. Inject the column with a gel fractionation buffer until the support is covered with the buffer up to 0.5 cm thickness.

4. Inject the GF buffer into the inlet tube until the support becomes wet.

5. Stir the prepared gel solution in Step 3 and pour it into the column. Add a GF buffer to fill the remaining step and place the lid on the column.

6. Fill the buffer reservoir with a GF buffer. Connect the reservoir to a pump with the help of a large tube.

7. Purge the pump with a GF buffer and attach the outlet from the pump to the inlet of the column. Open the column outlet and start the pump at the desired flow rate. Let the flow continue till the height of the gel bed becomes constant.

8. Once the gel bed height becomes constant, turn off the pump, close the column outlet, and remove the extension.

9. Assess the bed carefully for cracks, trapped air, and aggregates.

Prepare and Test the System

1. Calculate the amount of GF buffer necessary for the run and filter it plus a 50% excess through a 0.22-µm filter.

2. Assemble the GF system, placing the detector and recorder in line but leaving the column offline. Attach the buffer reservoir to the pump and purge it with a GF buffer.

3. Connect the outlet of the pump to the column *via* the injection valve, and run the system with a GF buffer at a flow rate set for separation.

4. Collect fractions with the fraction collector, and note the actual flow rate of the pump.

Evaluate the Column

1. Determine the void volume by running a coloured marker like 2 mg/ml Blue Dextran 2000 or 0.2 mg/ml vitamin B12 and obtain the void volume.

2. Calculate the total volume bypassing a total liquid volume by running a total liquid volume marker and obtaining the elution volume.

3. Calculate the separating volume of the column (Vi) by subtracting the void volume from the total volume (Vt – Vo).

4. Calculate the number of theoretical plates per column using the equation.

$N = 5.54m (V_r/W_h)2$

Where,

N = Number of plates per column

V_r = Elution volume of peak

W_h = Width of the peak at half peak height

5. Calculate the asymmetry factor of the peak according to the equation;

As = (b/a)

Where,

A = width of the leading part

b= width of the tailing part of the peak

<u>*Chromatograph the Sample*</u>

1. Dissolve the sample to be resolved in a GF buffer. Filter it through a 0.22 μm protein-compatible filter.

2. Open the outlet from the column, start the pump, and let two-bed volumes of the buffer pass through the column.

3. Load the sample buffer and adjust the column valve to the load condition.

4. Evaluate the collected fractions for purity.

Precautions

• While preparing any buffer involving tris, add the solvent first to prevent Tris from sticking at the bottom. Use HCl as a solvent while dissolving tris instead of water for better pH control.

• The sample volume should be less than 5% of the column volume to ensure elution as a sharp peak; the higher the ratio of sample to column volume, the lower the resolution. The sample buffer need not be the same as the chromatography buffer.

• The fractionation process is based on the particle sizes so any damage in the particle could lead to false results.

• The buffers and the matrix must be checked for trapped air; should be degassed before use. Trapped air can lead to poor resolution.

• Preparation of the gel from thin suspension and column packing in stages could result in a poorly packed column.

• Prevent disturbing the bed. Disturbing the bed will lead to uneven surfaces and hence uneven separation.

• The pore size of the matrix is decided according to the desired protein molecule; the rest tend to remain in the buffer.

Advantages

• Compatible with many types of solvents.

• Does not require particular protein attributes for elution.

• Suitable for the separation of similar species for example truncations and oligomers that might not be separated by any other chromatographic technique.

Disadvantages

• The outcome is severely dependent on column packaging.

• The non-specific interactions between protein and resin can lead to poor resolution.

• Complex protein solutions might lead to low resolutions.

• The loading volume of the sample must be low for proper resolution.

• Cannot be used for the initial purification of proteins when there are a lot of contaminations. The problems encountered in Gel Filtration Chromatography is summarized in Table **9.7**.

Table 9.7. Problems and their probable solutions in Gel Filtration Chromatography.

Problem	Solution
Poor resolution	- Minimize the sample volume. - Add salt to the buffer. A small amount of salt will help prevent nonspecific interactions between the protein and stationary phase. This will allow all protein molecules to run consistently over the length of the column. - Use a moderate flow rate. Running a column too quickly will not allow time for small molecules to enter the matrix pores, while a flow rate that is too slow allows more time for sample diffusion. - Ensure that sample and solvent viscosities are similar. Adjust sample conditions to approximate match that of the running buffer. - Adjust the column length. A column that is too short will not allow adequate protein separation. Alternatively, a column that is too long will lead to the diffusion of the protein sample. - Re-pack the column. Column packing can have a huge effect on protein resolution.
Protein does not elute	- Adjust the buffer conditions. - Change the stationary phase. - Change the flow rate.

Did you know?

• Are you a fan of programs like CSI and Law and Order? Well, in real life, chromatography is often used to investigate criminal cases like the ones featured on those shows.

• Artists use chromatography to identify the breakdown of colours for them to remake the colours they need.

CONCLUSION

With-tag and without-tag methods for protein purification are discussed. However, depending upon one's requirement and end application, the purification technique must be selected.

VIVA VOCE

1. What is the difference between half saturation and full saturation?

Ans: Half saturation and full saturation signify the amount of salt required for the precipitation of a protein. The amount of salt required for precipitation depends upon the molecular weight of a protein. There is generally an inverse relationship between the amount of salt required and the molecular weight of the protein. The proteins with a large molecular weight need less salt.

2. Since protein precipitation is dependent on which salt is used, which salt would precipitate protein at the lowest concentration of the salt solution: LiI, NaBr, K_2SO_4, and LiF.

Ans: K_2SO_4.

3. Protein Y was just discovered by scientists at a national lab. The scientists managed to purify the protein with some precipitate in their flask. However, their yields were very low, to solve the problem to extract the rest of the protein from the solution. They added 58 g of NaCl to 1 L of their protein solution to salt out the protein Y. After the addition of the NaCl, they noticed that the solution no longer had some of the previously precipitated protein. What is the reason for the disappearance of the precipitate?

Ans: The ionic strength of the solution after the addition of 58g of NaCl was about 1. In this case, the ionic strength was in the region where salting occurs. Hence the disappearance of the precipitate.

4. Apart from protein purification, what can dialysis be used for?

Ans: Hemodialysis: Removal of salts, and creatinine from blood in case of renal failure.

The concentration of protein in the dairy industry: Processing of cheese-whey to obtain whey protein concentrate and lactose-rich permeate.

Purifying Water: Use of removal of particulates and macromolecules from raw water to produce potable water.

5. What is the chief principle behind dialysis?

Ans: Dialysis works on the principles of the diffusion of solutes and ultrafiltration of fluid across a semipermeable membrane. Diffusion is a property of substances in water; substances in water tend to move from an area of high concentration to an area of low concentration.

6. What are the medical conditions for which dialysis is used as a treatment? Explain how?

Ans: Hemodialysis is a medical procedure designed to support the lost function of kidneys. In hemodialysis, the blood of the patient *via* a tube system, controlled by the dialysis machine, is passed through a filter. In this filter are hair-thin capillaries, which consist of a semipermeable membrane and are surrounded by a dialysis fluid. Toxic substances or blood salts can now pass through this semipermeable membrane so that in this way a concentration balance between the patient's blood and dialysis fluid for the said substances takes place and the blood is cleaned. The dialysis machine itself serves to monitor this process of transporting the blood through a system of pumps and to determine the composition of the dialysis fluid through which the exchange of substances and the reduction of the patient's blood volume is achieved.

7. Summarize the steps involved in Affinity Chromatography.

Ans: A. Binding: A complex solution containing the tagged protein is applied to the column and binds based on the affinity tag-matrix interaction B. Washing: Unnecessary proteins are washed away using buffers. C. Elution: Specifically bound protein is eluted from the column, typically by competitive binding of a similar molecule (*e.g.* histidine and imidazole), by cutting off the tag with a protease or by destabilization of the affinity tag-matrix interaction *e.g.* by a change of pH.

8. Explain non-specific elution.

Ans: Nonspecific elution can be done by using solvent or buffer changes, pH or ionic strength changes, the application of chaotropic reagents, temperature changes, reversible denaturation, and destruction of the spacer arm. Competition for suitable complementary protein is a type of specific elution.

9. Enlist some enzymes commonly used for tag removal from proteins.

Ans: Enterokinase, Aminopeptidase M, TEV Protease.

10. For gel filtration column chromatography, which two of the following have a linear relationship?

Ans: i. Amount of protein. ii. The relative elution volume of a protein (void volume over elution volume of a protein. iii. The logarithm of protein molecular mass. There is a linear relationship between the relative elution volume of a protein and the logarithm of its molecular mass.

11. Gel filtration chromatography is also known as? (2 names).

Ans: Size Exclusion Chromatography and Molecular Sieve Chromatography.

12. What do we aim to do when we say separation using gel filtration chromatography?

Ans: It can be used to either fractionate molecules and complexes in a sample into fractions with a particular size range, to remove all molecules larger than a particular size from the sample, or a combination of both operations.

13. What is gel permeation chromatography?

Ans: When an organic solvent is used as the mobile phase, the process is instead referred to as gel permeation. In this type of chromatography, the molecules are separated based on their hydrodynamic radius which is dependent on the size and shape of the protein.

REFERENCES

[1] Coligan, J.E. Current protocols in protein science. *Curr. Protoc. Protein Sci.,* **2001**.

[2] Arnau, J.; Lauritzen, C.; Petersen, G.E.; Pedersen, J. Current strategies for the use of affinity tags and tag removal for the purification of recombinant proteins. *Protein Expr. Purif.,* **2006**, *48*(1), 1-13. [http://dx.doi.org/10.1016/j.pep.2005.12.002] [PMID: 16427311]

[3] Harcum, S. Purification of protein solutions. In: *InBiologically inspired textiles*; Woodhead Publishing, **2008**; pp. 26-43. [http://dx.doi.org/10.1533/9781845695088.1.26]

[4] Grove, A.; Kushwaha, A.K.; Nguyen, K.H. *Determining the role of metal binding in protein cage assembly. In protein cages*; Humana Press: New York, NY, **2015**, pp. 91-100.

[5] Irvine, G.B. Determination of molecular size by size-exclusion chromatography *(gel filtration). Curr. Protoc. Cell Biol.,* **2001**, *Chapter 5*(1), 5. [PMID: 18228373]

CHAPTER 10

Introduction to Nucleic Acids

Sai Joshi[1,*], **Sinchan Hait**[2] and **Ahiree Ghosal**[3]

[1] *CSIR-NCL, Dr. Homi Bhabha Road, Pune - 411 008, India*

[2] *Department of Chemical Engineering, BITS-Pilani K.K. Birla Goa Campus, Goa-403726, India*

[3] *Zydus School for Excellence, Ahmedabad, Gujarat, India*

Abstract: Nucleic acids are biomolecules which carry all the genetic information in the form of DNA and RNA. The nucleic acids are composed of primarily a 5-carbon sugar, a phosphate and a nitrogenous basis. These nitrogenous bases can be adenine, guanine, thymine, cytosine and uracil. Nucleic acids can be measured qualitatively by UV absorbance, diphenylamine method/orcinol method, fluorescence dyes and capillary electrophoresis. It can be measured qualitatively by agarose gel electrophoresis.

Keywords: Capillary electrophoresis, Diphenylamine method, DNA, Fluorescence dyes, Nucelic acid, Orcinol method, RNA, UV absorbance.

INTRODUCTION

Nucleic acids are polymers of nucleotides, which are monomers made of three components: a 5-carbon sugar, a phosphate group, and a nitrogenous base (Fig. **10.1**) for nucleic acid composition. They are biomolecules which are also known as the energy currency of metabolism transactions due to the following functions [1 - 3]:

(a) They are an important chemical linkage between cellular response to hormones and other extracellular stimuli.

(b) They are the molecular repository of genetic information passed down from one generation to another, which is a fundamental property of life.

(c) The structural characteristics of proteins and, ultimately, every biomolecule is dependent on information programmed by the nucleic acid sequence of the cell. Nucleic acids comprise 3 components [1, 2]:

* **Corresponding author Sai Joshi:** CSIR-NCL, Dr. Homi Bhabha Road, Pune - 411 008, India; E-mail: saijoshi1997@gmail.com

Pamela Jha (Ed.)

5-carbon sugar, which is usually ribose, a phosphate group and a nitrogenous base.

Nitrogenous bases are of two kinds: (a) Pyrimidine: It is a heterocyclic aromatic ring containing two nitrogen atoms at the C1 and C3 positions of a 6-carbon ring.

(b) Purine: It is a pyrimidine ring with an imidazole ring.

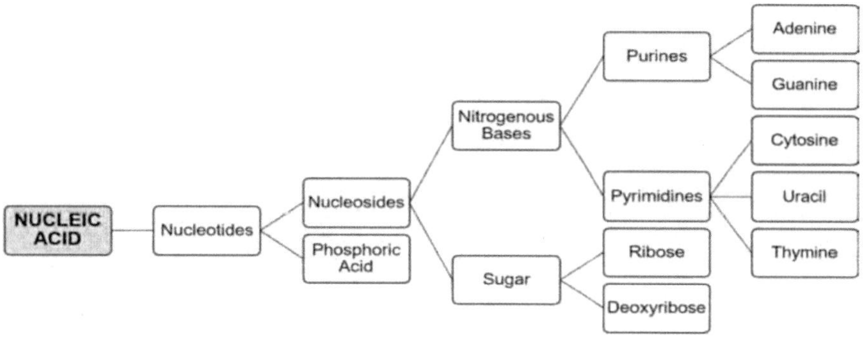

Fig. (10.1). Nucleic acid structure.

Nucleic acids can be further divided into [1 - 3]:

Deoxyribonucleic Acid (DNA)

The synthesis of functional biological products is largely dependent on cellular DNA, referred to as a **gene**. Cells possess thousands of genes and therefore their size tends to be very large.

Functions of DNA

Storage and transmission of genetic information in the form of genes.

Components of DNA

5 carbon sugar: Deoxyribose

Nitrogenous bases: Adenine (A), Guanine (G), Thymine (T), Cytosine (C)

Phosphate group

RNA: Ribonucleic Acid

RNA sequence is dependent on DNA sequence, and it has a broader range of functions than DNA. They are classified as:

Ribosomal RNA (rRNA)

They are a component of ribosomes, which play an important role in protein synthesis.

Messenger RNA (mRNA)

They carry genetic information from a gene which is to be translated to protein to the ribosome.

Transfer RNA (tRNA)

Translate data in mRNA into a specific amino acid sequence.

Components of RNA

5 carbon ring: Ribose

Nitrogenous bases: Adenine (A), Guanine (G), Uracil (U), Cytosine (C)

Phosphate group

Nucleic acids have two kinds of pentoses: DNA contains 2'- deoxy-D-ribose, whereas RNA contains D-ribose pentose sugar. Depending on this property of nucleic acids, the classification of DNA or RNA is established. Although DNA and RNA contain the major bases (A, G, C, T (DNA), U (RNA)), however, some minor bases have been identified. These are:

(a) In DNA, usually, the methylated forms of nucleic acid bases are found. *E.g.*: 5-methylcytidine occurs in the DNA of animals and higher plants.

(b) In viral DNA, hydroxymethylated or glycosylated forms of nucleic acid bases are found. *E.g.*: 5-hydroxymethylcytosine in the DNA of animals and bacteria infected by bacteriophages.

These altered forms of DNA play a role in the regulation or protection of the genetic sequence.

(c) Minor bases are also found in RNAs, especially tRNAs. *E.g.*: Inosine contains the base hypoxanthine.

The long chain nucleotides are formed by the phosphodiester linkage between the 5' - phosphate group of one nucleotide binds to the 3'- hydroxyl group.

The purines and pyrimidines that are common in DNA and RNA are aromatic molecules, a property with important consequences for the structure, electron distribution, and light absorption of nucleic acids. Electron delocalization among atoms in the ring gives most of the bonds partial double-bond character. All nucleotide bases absorb UV light, and nucleic acids are characterized by a strong absorption at wavelengths near 260 nm.

The Structure of Nucleic Acids

The structure of nucleic acids can be categorized as [1, 2]:

(a) Primary: Covalent structure and the nucleotide sequence

(b) Secondary: Stable structure taken up by nucleotides

(c) Tertiary: Folding of large chromosomes within chromatin

The Establishment of DNA as Genetic Information

Isolation of DNA

The isolation and characterization of DNA were performed by Friedrich Miescher in 1868. He called this substance "nuclein".

Evidence that DNA is the Genetic Material

In 1944, Oswald T. Avery, Colin MacLeod, and Maclyn McCarty found that DNA was extracted from virulent *Streptococcus pneumoniae* when injected into a non-virulent strain of *Streptococcus pneumoniae,* the non-virulent strain transformed into a virulent one. The conclusion was affirmed as DNA carried the genetic information for the virulence of the bacterium.

However, it was not widely accepted as researchers thought the structure of nucleic acids was simple and it was not a complex molecule like proteins.

DNA Accepted as the Genetic Material

In 1952, Alfred D. Hershey and Martha Chase studied the virus which infects bacteria - bacteriophage - which had radioactively labelled DNA or protein. They found that bacteriophage transmits radioactively labelled DNA to the bacteria, which removed the doubt that DNA, not proteins, carried genetic information.

Erwin Chargaff Rules

In the 1940s, Erwin Charfgaff and his colleagues found that four nucleotides have a characteristic binding pattern which can be explained with the help of ratios of nucleotides in the DNA.

This led to Chargaff's rules:

The base composition of DNA generally varies from one species to another.

DNA specimens isolated from different tissues of the same species have the same base composition.

The base composition of DNA in a given species does not change with an organism's age, nutritional state, or changing environment.

In all cellular DNAs, regardless of the species, the number of adenosine residues is equal to the number of thymidine residues (that is, **A = T**), and the number of guanosine residues is equal to the number of cytidine residues (**G = C**). From these relationships, it follows that the sum of the purine residues equals the sum of the pyrimidine residues; that is, **A+G = T+C.**

DNA X-ray Fiffraction Pattern

Rosalind Franklin and Maurice Wilkins used the powerful method of X-ray diffraction to analyze DNA fibres. They showed in the early 1950s that DNA produces a characteristic X-ray diffraction pattern. From this pattern, it was deduced that DNA molecules are helical with two periodicities along their long axis, a primary one of 3.4 Å and a secondary one of 34 Å.

Qualitative Analysis of Nucleic Acids

The method chosen for the qualitative analysis of nucleic acids is Agarose gel electrophoresis:

Electrophoresis separates charged components of a heterogeneous sample by differences in their migration rates in an electric field. The vast majority of important biological molecules, including amino acids, peptides, proteins, nucleotides, and nucleic acids, contain ionizable groups and, therefore, exist in solution as either cation (+) or anions (-). Under the influence of an electric field, these charged particles migrate either to the cathode or to the anode, depending on their net charge.

To prepare agarose gels, dry agarose is suspended in an aqueous buffer, and then the mixture is boiled until a clear solution is obtained. Once poured and cooled to room temperature, this forms a rigid gel. Large pores are formed using lower concentrations of agarose, while smaller pores are formed using higher concentrations.

Besides agarose, there are other important components necessary for carrying out electrophoresis. Several buffers have been recommended for the electrophoresis of DNA. TAE (Tris-acetate-EDTA) and TBE (Tris-borate-EDTA) are the most commonly used. Due to differences in ionic strength, DNA fragments migrate at different rates in these two buffers. Besides establishing pH, buffers also contribute to ions needed for conductivity.

DNA gel loading dye contains Bromophenol blue and glycerol. Using bromophenol blue, we can measure how far DNA migrates in a gel while it runs so we can measure the distance it covers. In addition, DNA settles at the bottom of each well as a result of the density of the glycerol present in the DNA loading dye.

A fluorescent dye ethidium bromide is used for tagging DNA. Since DNA has no colour of its own, EtBr intercalates with the bases to give fluorescent colour under UV light.

Quantitative Analysis of Nucleic Acids

In choosing your DNA quantification method, there are many things to consider: cost, time, equipment, and expected DNA concentration.

There are 4 common methods:

1. UV absorbance

2. Diphenylamine method/Orcinol method

3. Fluorescence dyes

4. Capillary electrophoresis,

in addition to digital droplet digital PCR, which are precise, reproducible and absolute without the use of a standard curve.

Spectrophotometric Analysis

Spectrophotometry can be used quantitatively since there is a relationship between the amount of light absorbed by a solution and the concentration of the nucleic acids in the solution. For many substances, the amount of light absorbed depends on the solute concentration and the length of the light path. This relationship is known as *Beer-Lambert's Law* and is expressed as:

$$A = \alpha c l$$

where A is the absorbance, the constant α is the absorption coefficient ($cm^2/\mu g$), c is the solute concentration ($\mu g/ml$), and l is the length of the light path (cm). A plot of absorbance versus concentration gives a straight line with slope α.

Fluorometric Methods for Nucleic Acid Quantitation

Fluorometric methods are widely regarded as some of the most sensitive methods available for nucleic acid quantitation. These methods use fluorescent dyes that intercalate and bind nucleic acid grooves, bind DNA or RNA non-specifically, or selectively bind nucleic material. Differences in spectral characteristics of nucleic acid-bound fluorophores allow sample concentrations to then be determined.

Each nucleic acid dye has a specific excitation and emission wavelength. The DNA or RNA sample is measured using a fluorometer, and nucleic acid concentrations are then calculated by comparing the fluorescence emission of the sample to a fluorescence curve generated using standards of known nucleic acid concentration. Different sample types (genomic DNA, plasmid DNA, *etc.*) require their standard curves. As with absorbance methods, calculated concentrations must be corrected for dilution factor as required.

Considerations when selecting a fluorescent nucleic acid dye for quantitation:

Specificity

Different dyes are optimized to measure dsDNA, ssDNA, RNA, or small RNAs (*e.g.*, miRNA) based on their mechanism of binding and spectral properties - make sure you select a fluorescent dye that binds to your target of interest.

Sensitivity

Highly sensitive fluorescent dyes allow you to use fewer samples or measure very low concentrations of nucleic acid.

Dynamic Range

Some reagents can be used to detect nucleic acid at low concentrations, but may lose linearity at higher concentrations; the reagent to be used has to be suitable for the expected concentration range of your samples.

Sample Volume

If your sample volume is limited, pay attention to the sample volume required by the reagent or kit.

Sample Format

Some reagents and kits have been optimized for measurement in single tubes or cuvettes, while others are intended for use in microplates for higher throughput

Instrumentation Filters

Check if your fluorometer has suitable excitation and emission filters for the selected dye.

Capillary Electrophoresis

Capillary electrophoresis (CE), although comparatively new, offers the advantages of an instrumental approach to electrophoresis. The capillary format enables the generation of electric fields at least 50-fold more powerful than those of traditional slab gels, which leads to rapid separations on the order of minutes. Because of its great potential, its suitability for biological mixtures, and its ability to handle precisely nanoliter quantities of material, CE is fast becoming one of the most widely used techniques for analytical separation.

CONCLUSION

Nucleic acids are macromolecules essential for all life forms. It can be RNA or DNA based on nitrogenous bases. It can be detected in a sample using agarose gel electrophoresis. Nucleic acids can be measured qualitatively by agarose gel electrophoresis and quantitatively by UV absorbance, diphenylamine method/orcinol method, fluorescence dyes and capillary electrophoresis.

VIVA VOCE

1. What are the major components of nucleic acid?

Ans. Components of nucleic acids are 5-carbon sugar which is usually ribose, a phosphate group and a nitrogenous base (*i.e.*, purine and pyrimidine).

2. Explain the different structures of nucleic acids.

Ans. The different structures of nucleic acids are as follows:

Primary: Covalent structure and the nucleotide sequence.

Secondary: Stable structure taken up by nucleotides.

Tertiary: Folding of large chromosomes within chromatin.

3. Which experiment confirmed DNA as a genetic material?

Ans. Alfred D. Hershey and Martha Chase's experiment confirms that DNA acts as a genetic material.

4. What is the principle of agarose gel electrophoresis?

Ans. The DNA molecule is separated based on the charge and size under the applied electric field, depending upon the pore size of the gel.

REFERENCES

[1] Watson, J.D.; Crick, F.H.C. Molecular structure of nucleic acids; a structure for deoxyribose nucleic acid. *Nature,* **1953**, *171*(4356), 737-738. [http://dx.doi.org/10.1038/171737a0] [PMID: 13054692]

[2] Watson, JD; Crick, FH *A structure for deoxyribose nucleic acid*; Die Entdeckung der Doppelhelix, **1953**, pp. 141-143. [http://dx.doi.org/10.1007/978-3-662-47150-0_4]

[3] Eschenmoser, A. Chemical etiology of nucleic acid structure. *Science,* **1999**, *284*(5423), 2118-2124. [http://dx.doi.org/10.1126/science.284.5423.2118] [PMID: 10381870]

CHAPTER 11

Qualitative Tests for Nucleic Acid Detection

Anshika Sah[1], **Mrittika Sarkar**[2], **Sinchan Hait**[3], **Sai Joshi**[4,*] and **Ahiree Ghosal**[5]

[1] *Environmental Studies and Resource Management, TERI School of Advanced Studies, New Delhi - 110 070, India*

[2] *Department of Biomedical Engineering, University of Michigan, Michigan, USA*

[3] *Department of Chemical Engineering, BITS-Pilani K.K. Birla Goa Campus, Goa-403726, India*

[4] *CSIR-NCL, Dr. Homi Bhabha Road, Pune - 411 008, India*

[5] *Zydus School for Excellence, Ahmedabad, Gujarat, India*

Abstract: Qualitative tests of nucleic acids are important to detect the presence or absence of nucleic acid in the given sample. It can be done by performing agarose gel electrophoresis. Nucleic acids are negatively charged and hence can be attracted towards a positive charge in an electric field. This phenomenon is used in agarose gel electrophoresis to detect nucleic acids. Based on the mass of nucleic acids, different bands are formed in the gel, which can be visualized using fluorescent dyes.

Keywords: Agarose gel, DNA, Nucleic acids, Negative charge, RNA.

INTRODUCTION

Detection of nucleic acid can be done efficiently using agarose gel electrophoresis. The pores of the agarose gel allow the sample to flow through it. The size of the pores determines the sieving property of the sample and, therefore the size of nucleic acid which can pass through the gel [1]. Nucleic acid with different sizes has different molecular masses, which are easily differentiated in the agarose gel.

AGAROSE GEL ELECTROPHORESIS

Aim

To test the presence of DNA/RNA in the sample by agarose gel electrophoresis.

* **Corresponding author Sai Joshi:** CSIR-NCL, Dr. Homi Bhabha Road, Pune - 411 008, India;
E-mail: saijoshi1997@gmail.com

Pamela Jha (Ed.)

Principle

Having phosphate backbones, DNA (and RNA) molecules are negatively charged, which causes them to migrate toward positively charged anodes when they are exposed to an electric field. The mobility of NA molecules is influenced by the applied voltage. Sample migration is faster at higher voltages within a range. Large molecules move more slowly due to frictional drag and less efficiently pass through pore spaces in gels than smaller molecules [1 - 3].

The rate of migration (v) of DNA/RNA is dependent on electrophoretic mobility, given by the following equation:

$$v=(qE)/f$$

Where, v= velocity, q= charge on particle, E= electric field (volts/cm), f= frictional coefficient

The bands are visualized with fluorescent dyes, ethidium bromide or SYBR Green which intercalates between nucleic acid bases.

Range: Separates DNA fragments of size ranging from 100 bp to 25 kb.

Reagents and Materials

- **Equipment**

- Electrophoresis apparatus: agarose gel, an electrophoretic chamber with a cathode (negative terminal) at one end and an anode (positive terminal) at the other end.

- Casting tray

- Well combs

- Voltage source

- UV light source

- Microwave

- **Reagents**

DNA Electrophoresis

- TAE- 50X (Tris-base: 242 g Acetate (100% acetic acid): 57.1 ml EDTA: 100 ml 0.5M sodium EDTA. Add dH2O up to one litre. To make 1x TAE from this stock, dilute 20ml of stock into 980 ml of dt water)

• Agarose (1%)

• Ethidium bromide (10 mg/mL)

RNA Electrophoresis

• **Gel loading buffer II** - Denaturing PAGE

• A 1-2X solution of 95% Formamide, 10 ml 10X MOPS running buffer, 18mM EDTA, and 0.025% SDS, Xylene Cyanol, and Bromophenol Blue.

• **Gel loading solution** - All-purpose, native agarose

• A 10X solution of 40% Sucrose, 0.17% Xylene Cyanol, and 0.17% Bromophenol Blue

Procedure

1. Prepare a standard 1% agarose gel.

2. Loading samples and running an agarose gel:

DNA

• Before the agarose gel solidifies, add the ethidium bromide (EtBr) at a concentration of approximately 0.2-0.5 g/mL (about 2-3 mL of stock solution per 100 mL of gel).

• Add a loading buffer to each of your DNA samples.

• Upon solidifying, place the gel into the electrophoresis unit.

• Fill the gel box with 1xTAE (or TBE) until the gel is covered.

• In the first gel lane, load the DNA ladder.

• Carefully load your samples into the additional wells of the gel.

• Run the gel at ~50–70 mA of current, until the dye line is approximately 75-80% of the way down the gel. In general, a gel run lasts between one and 1.5 hours, depending on the concentration and voltage of the gel.

• You must turn off the power, disconnect the electrodes, and carefully remove the gel from the gel box after the disconnection of the electrodes.

RNA

• To 1-3 g RNA, add 0.5-3 x volume of Formaldehyde Load Dye. A final concentration of 10 g/ml of ethidium bromide can be added to the Formaldehyde Load Dye.

• Heat denatures samples at 65-70°C for 5-15 min.

• Load the gel and electrophorese at 5-6 V/cm until the bromophenol blue (the faster-migrating dye) has migrated at least 2-3 cm into the gel, or as far as 2/3 the length of the gel.

3. Visualize the gel on a UV transilluminator/Gel Doc.

4. Analyze the gel:

• Determine the sample DNA size by comparing it with DNA ladder fragments.

Precautions

• The agarose must not be over-boiled to prevent buffers from evaporating and altering the final agarose content.

• Pour the agarose slowly into the gel tray to prevent bubbles from disrupting the gel. (Tip: Bubbles can be removed with a pipette tip by pushing them away from the comb or towards the sides or edges of the gel.)

• EtBr is a known mutagen; take proper precautions while handling it.

• The gel agarose percentage can be altered to get a better separation if sample DAN have too many bands that are of the same size. An agarose gel with a higher percentage will help to separate smaller bands, and a gel with a lower percentage will help to separate larger bands.

• As RNA can be easily denatured by RNAse, before preparing the RNA samples, wash the buffer recirculation pump and hoses by circulating with 0.2N NaOH for 30 minutes and then rinsing with ddH_2O.

Interpretation of sample gel images and troubleshooting (Fig. **11.1**).

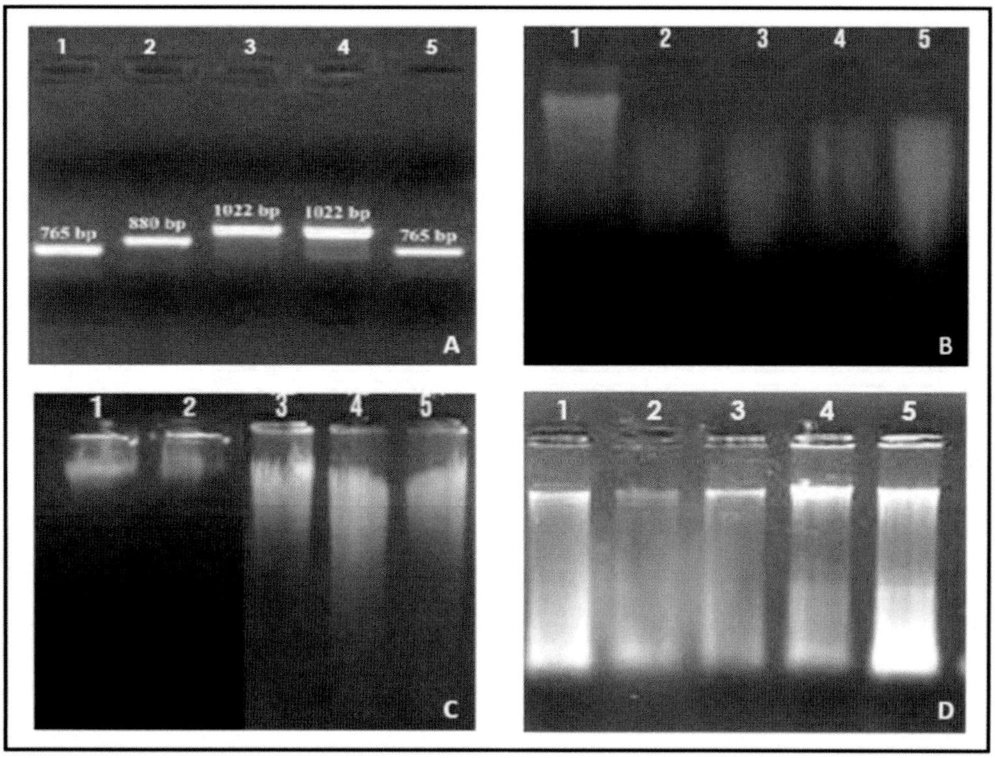

Fig. (11.1). Gel images.

Gel A

The image represents a good gel where bands observed in all the lanes (Lane 1 to Lane 5) are single, prominent and distinct.

Gel B

Comments: Lane 1: the DNA has not come out from the gel but has not migrated properly and the smear above the DNA indicates contamination of the RNA. Lane 2 to 5: All other DNAs are degraded.

Troubleshooting

- RNA contamination can be removed by adding Rnase.

Gel C

Source: gDNA

Comments: From Lanes 1 to 5, it can be observed that in each well, DNA is trying to come out of the well, but some DNA remains inside the well.

Firstly, the wells might be broken during sample loading (for lanes 3 to 5), and secondly, the air bubbles were formed during the gel casting.

The gel may also be affected by incorrect placement of the comb or by the comb being disturbed during removal. These factors prevent gDNA from flowing from the well.

As noted above, contamination of the DNA band by RNA or protein (in wells 3,4,5) indicates inadequate DNA extraction and poor gel preparation further suggests a poor DNA extraction and poor gel preparation.

Troubleshooting

• DNA isolation protocol needs to be optimized for higher yield and less RNA and protein contamination.

• RNA contamination can be removed by adding RNase A.

Gel D

Comments: The bands are not distinct and have formed a smear.

Troubleshooting

• Higher concentration of DNA can cause restriction in the mobility of DNA, which may get condensed and get aggregated, reducing its transfer and movement through the gel lane. This may have resulted in the smear formation.

• In addition, the sample should be treated with RNase enzyme to avoid RNA contamination.

NUCLEIC ACID QUALITY AND QUANTITY ESTIMATION BY NANODROP SPECTROMETRY

Principle

Nanodrop is a modified form of UV spectrophotometer which works on the similar principle of conventional spectrophotometer but has many more features. This instrument works by combining fibre optic technology and natural surface tension properties to capture and retain very small amounts of sample independent of traditional containment apparatus such as capillaries or cuvettes. Because of its use of sample surface tension, it requires 1μL- 2μL of sample for analysis. This reduction of the volume of sample required for analysis facilitates the incorporation of additional quality control steps, which increases efficiency and leads to great downstream results. Moreover, this system also employs shorter path lengths, which measure a broad range of nucleic acid concentration, thus, eliminating the requirement to perform dilutions. During the working of the instrument, the sample is retained on an optic fibre, after which it is assessed to an absorbance range of 220nm to 750nm to determine the quality and quantity of nucleic acids. In this technique, first, a direct A260 absorbance method is used, followed by a demonstration of a fluorescence-based method which enables fluorescence reactions of reduced-volume a fluoro spectrometer. This results in the assessment of nucleic acid concentrations ranging from 1 pg/µl to 15,000 ng/µl [4, 5]. Furthermore, it also calculates the contaminants concentration of the sample through the A260/A280 and A260/A230 ratios, which can be seen very prominently on the entire graph of the absorbance spectrum of the sample displayed [4, 5].

Requirements

Equipment- NanoDrop 2000c Spectrophotometer, Micropipettes

Chemicals- TE buffer, unknown sample

Others- Deionized water (D/W), lint-free lab wipe

Procedure

1. Pipette 2 to 3 µL of clean D/W onto the lower optical surface of the spectrophotometer to clean the upper and lower optical surfaces of the spectrophotometer.

2. Select the Nucleic Acid application on the NanoDrop software.

3. Perform a blank measurement by dispensing 1 µl of TE buffer onto the lower optical surface.

4. Select "Blank" in the nucleic acid application and lower the lever arm.

5. Clean both optical surfaces.

6. Choose the appropriate constant options for the sample that is to be measured.

7. Now, onto the lower optical pedestal, dispense 1µl of sample and close the lever arm.

8. Click on "Measure" in the application software.

9. The software automatical calculates the purity ratio and concentration of nucleic acid (Table **11.1**).

Table 11.1. Constants of DNA and RNA are mentioned.

Sample Type	Option	Constant for Calculating the Concentration
ssDNA	DNA-33	33
dsDNA	DNA-50	50
RNA	RNA-40	40
Oligo	Custom	15-150

Results

Example of a typical spectrum of a sample containing nucleic acid and contaminants through the NanoDrop Technique.

A typical nucleic acid sample usually has a very characteristic spectrum or profile, as shown below in the Fig. (**11.2**).

If the sample consists of any contaminants, the spectrum obtained will be very different from the characteristic profile. Shifts in the peaks and troughs of the spectra are observed. *E.g.*, in Fig. (**11.3**) given below, a comparison of spectra of nucleic acid samples is shown with and without contaminants. In this figure, it is observed that samples with contaminants (phenol & guanidine) have relatively less absorbance as compared to the sample without contaminants at a wavelength of 260 nm.

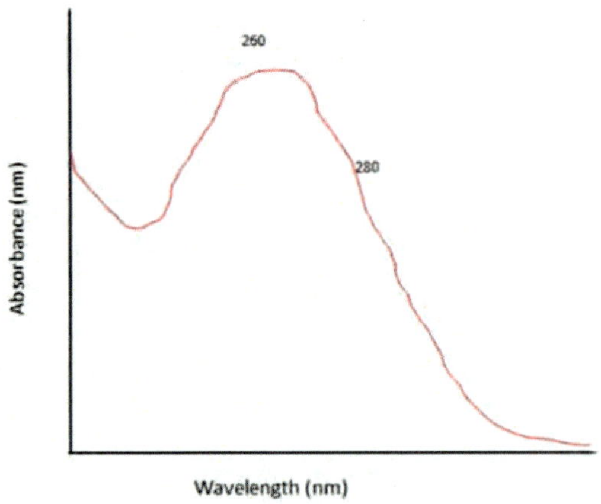

Fig. (11.2). Example of the spectrum of a sample containing nucleic acid.

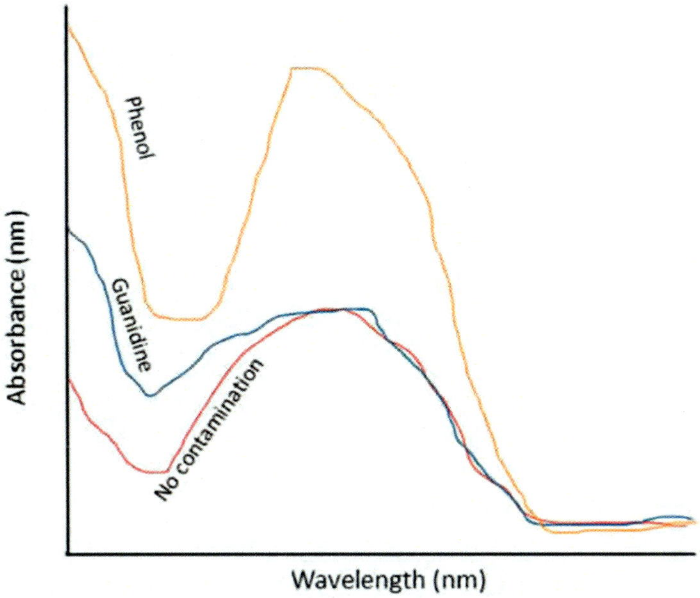

Fig. (11.3). Comparison of a shift of peaks and spectra nucleic acid sample without contamination (A=Normal) and with 2 contaminants (B= Guanidine & C= Phenol).

Furthermore, the quality of the sample assessed is also determined by A260/A280 and A260/A230 ratios, which tell about the protein and RNA contamination, respectively.

Pure nucleic acids yield A260/A280 ratio for DNA is ~ 1.8.

Pure nucleic acids yield A260/A280 ratio for RNA is ~ 2.0.

Pure nucleic acids yield A260/A230 ratio for DNA 1.8-2.2.

Did you know?

• Nucleic acids were first discovered by Friedrich Miescher. It was first seen in the nuclei of white blood cells. It was named the nucleus as it was obtained from the nuclei.

• Altman discovered the acidic nature of nucleic acids.

• Mathaei, Nirenburg, and Khorana synthesized nucleotides *in vitro* as part of their exercise to understand the genetic code dictionary and unravel the exact mechanism of translation of the language of nucleic acids into that of proteins.

• Humans share 60% of genes with fruit flies, and 2/3 of those genes are known to be involved in cancer.

• The closest invertebrate genetic relative of humans is a small creature known as the star ascidian or golden star tunicate.

CONCLUSION

Nucleic acids are the basis of genetic information in living organisms. The presence of genetic material in a given sample is a piece of important information. The DNA and RNA in the given sample can be detected using agarose gel electrophoresis. In the agarose gel, bands are formed of nucleic acids which can be visualized using fluorescent dye.

VIVA VOCE

1) What is the difference between agar and agarose?

Ans. Agar is a gelatinous substance, while agarose is a linear polymer. Both are polysaccharide products obtained from red algae and seaweed, respectively. The structure of agarose is a linear polymer consisting of alternating D-galactose and 3,6-anhydro-L-galactose units. Agar differs from agarose because it is composed of both agaropectin and agarose. Agarose is isolated from the seaweed genera *Gelidium* and *Gracilaria* and consists of repeated agarobiose (L- and D-galactose) subunits.

2) Suggest a different method for DNA separation besides agarose gel electrophoresis.

Ans. Sucrose density gradient centrifugation (only provides an approximation of size).

3) What are the factors that influence the rate of the migration of nucleic acid molecules?

Ans. The rate of migration of a DNA molecule through a gel is determined by the following:

a) size of the DNA molecule; b) agarose concentration; c) DNA conformation; d) voltage applied, e) presence of ethidium bromide, f) type of agarose and g) electrophoresis buffer.

4) What are the absorbance maxima of nucleic acids?

Ans. According to the Beer-Lambert law, the amount of light absorbed at 260 nm is proportional to the concentration of nucleic acid in the solution. For dsDNA, RNA, and ssDNA, extinction coefficients were determined using a 10 mm path length and allowed conversion factors to be calculated without a standard curve. The conversion factors are: 50 g/ml for dsDNA, 40 g/ml for RNA, and 33 g/ml for ssDNA.

For example, a dsDNA sample with $A_{260} = 1$ will have a concentration of 50 ng/µl. The $A_{260/280}$ provides insight regarding the type of nucleic acid present (dsDNA or RNA) as well as a rough indication of the level of purity. In buffered solutions, pure dsDNA has an $A_{260/280}$ of 1.85–1.88 and pure RNA has a ratio of around 2.1.

5) What is the role of loading dyes in the technique?

Ans. In-gel electrophoresis, loading dyes serve three main purposes. 1) They enable the sample to sink due to high density. 2) Dye provides colour and simplifies the loading of the gel. 3) Dyes move through the gel at standard rates, enabling the distance of migration of DNA fragments to be estimated.

6) How do the different forms of DNA move through the gel?

Ans. Different forms of DNA move through the gel at different rates. For example, Supercoiled plasmid DNA, because of its compact conformation, moves through the gel fastest, followed by a linear DNA fragment of the same size, with the open circular form travelling the slowest.

7) What are the commonly used denaturing systems used for RNA gel electrophoresis?

Ans. Formaldehyde, formaldehyde/formamide, and glyoxal, in addition to DMSO, are commonly used to denature RNA in agarose gels.

8) Can TAE/TBE buffers be used for RNA gel electrophoresis?

Ans. When cellular RNA integrity is being checked, no sample denaturation is necessary, TAE or TBE is suitable.

But, two of the most common buffer systems for the electrophoresis of denatured RNA include a phosphate buffer for glyoxal/DMSO denatured RNA and a MOPS buffer for formaldehyde or formamide.

9) Are the denaturants pH dependent?

Ans. pH is a major factor in the denaturants' activity. As a result of pH gradients generated during electrophoresis, agarose gel may melt (hydrolyze). When the buffer pH rises above pH 8.0, glyoxal dissociates from RNA, causing the RNA to renature and migrate in an unpredictable direction.

By recirculating the buffer, this problem can be avoided.

10) How does EtBr stain the DNA strands?

Ans. DNA is most often stained using EtBr in agarose gels. EtBr interacts with DNA by intercalating itself in the molecule in a concentration-dependent manner. Ethidium molecules absorb ultraviolet light, which leads to the excitation of electrons in the aromatic rings. Electrons revert to their ground state and release energy.

11) Name a few alternative dyes that can be used instead of EtBr. Are they efficient alternatives for EtBr?

Ans. For DNA staining in agarose gels, alternative stains include SYBR Gold, SYBR Green, Crystal Violet and Methyl Blue. When using methyl blue and crystal violet to visualize DNA bands, the gel does not need to be exposed to ultraviolet light, reducing the chance of mutation when recovering the DNA fragment.

Yet their sensitivity is still lower than EtBr. In comparison to EtBr, both SYBR gold and SYBR green are highly sensitive dyes that are UV dependent with a low toxicity level, but they are more expensive. All of the other dyes are either incompatible with the gel or do not work well when added directly to it, which is why the electrophoresis gel will have to be post-stained after it has been electrophoresed.

EtBr remains a dye of choice for many researchers due to its ease of use, sensitivity, and cost-effectiveness when handled according to safety protocols.

12) What is the detection range for Nanodrop?

Ans. Nanodrop can detect DNA concentration in a range of 1pg/µl - 15,000ng/µl.

13) Which of the following reagents would be useful for the visualization of DNA restriction fragments that have been separated by electrophoresis in an agarose gel and remain in the wet gel?

A. Ethidium bromide

B. [Alpha-32 P] ATP

C. Diphenylamine

D. 32Pi

Ans. Ethidium bromide.

14) Why is nanodrop spectrometry considered to be an efficient method?

Ans. Nanodrop spectrometry requires only 1μL- 2μL of the sample. By requiring a smaller quantity of samples for analysis, it is possible to incorporate more quality control procedures, which improves efficiency. Also, the elimination of the dilution process makes it more efficient.

15) How to identify the presence of contaminants in the sample in Nanodrop Spectrometry?

Ans. The sample with contaminants will have less absorbance as compared to the sample without contaminants and the spectrum obtained will be very different from the characteristic profile. Further, A260/A280 and A260/A230 ratios tell about the protein and RNA contamination, respectively.

16) A DNA sample and a protein sample are mixed. The mixture is equally divided into A and B. NaCl concentrations are 0.01M and 1M in mixture A and B, respectively. In sample A, the CD spectrum of DNA is changed, but that of protein is not. In sample B, neither is changed. The CD spectra of pure DNA in 0.01M NaCl and 1M NaCl are the same. What could be the possible explanations?

P. Protein binds to DNA in 0.01M but not in 1M NaCl.

Q. At 0.01M NaCl, protein induces a change in the helicity of the DNA

A. Only P is correct

B. Only Q is correct

C. Both P and Q are correct

D. Both P and Q are incorrect

Ans. (A) only P is correct.

REFERENCES

[1] Lee, P.Y.; Costumbrado, J.; Hsu, C.Y.; Kim, Y.H. Agarose gel electrophoresis for the separation of DNA fragments. *J. Vis. Exp.,* **2012**, (62), e3923.
[PMID: 22546956]

[2] Sambrook, J.; Russell, D.W.; Sambrook, J. *The condensed protocols: From molecular cloning: A laboratory manual*; Cold spring harbor laboratory press: Cold Spring Harbor, NY, **2006**.

[3] Krupyanko, V.I. Experimental biochemistry, and company.*Process Biochem*; Switzer, R.L.; Garrity, L.F.; Freeman, W.H., Eds.; Elsevier Science: New York, **2000**, 4, p. (36)379.
[http://dx.doi.org/10.1016/S0032-9592(00)00214-4]

[4] Desjardins, P.; Conklin, D. NanoDrop microvolume quantitation of nucleic acids. *J. Vis. Exp.,* **2010**, (45), e2565. [Journal of Visualized Experiments].
[PMID: 21189466]

[5] Shen, C. H. Detection and analysis of nucleic acids. *Diagnostic Molecular Biology,* **2019**, 167-185.

CHAPTER 12

Quantitative Tests for Nucleic Acid Detection

Mrittika Sarkar[1], **Sinchan Hait**[2], **Sai Joshi**[3,*] and **Revati Supekar**[4]

[1] *Department of Biomedical Engineering, University of Michigan, Michigan, USA*

[2] *Department of Chemical Engineering, BITS-Pilani K.K. Birla Goa Campus, Goa-403726, India*

[3] *CSIR-NCL, Dr. Homi Bhabha Road, Pune - 411 008, India*

[4] *Biotechnology, Dpartment of Management Studies, DY Patil University, Pune, India*

Abstract: Nucleic acid quantification is essential in experiments associated with genetic engineering. Over periods, several techniques have been developed to quantify nucleic acids. UV-Vis spectrophotometer is the simplest technique where it is associated with absorbance, which is proportional to the concentration of the sample. There are some colourimetric methods, such as the diphenylamine (DPA) method, and the orcinol method, where nucleic acid reacts with diphenylamine and forms a bright blue colour complex, and orcinol reacts and forms a green colour complex. Apart from that, some advanced techniques, such as fluorescence techniques, are used for quantification. In this chapter, each of these methods is explained in detail.

Keywords: Diphenylamine (DPA) method, Fluorescence, Nucleic acid, Orcinol method, UV-Vis spectrophotometer.

INTRODUCTION

Besides the qualitative analysis of nucleic acids, quantification of the same is an important step. Over a while, several techniques have been developed, however, there are some standard assays available which are performed routinely. UV-Vis spectrophotometric analysis of quantification is the simplest one to perform [1] [3]. In the diphenylamine (DPA) method, the diphenylamine reacts with diphenylamine and forms a bright blue colour complex which is measured at 600nm [2]. Whereas, in the orcinol method, orcinol reacts with orcinol and forms a green colour complex, and the absorbance is measured at 660nm [1, 3]. In advanced techniques, fluorescence dyes are used, which bind to the nucleus acids and upon excitation, they emit some light which is then quantified to determine the amount of nucleic acid present in the sample [4 - 6].

* **Corresponding author Sai Joshi:** CSIR-NCL, Dr. Homi Bhabha Road, Pune - 411 008, India;
E-mail: saijoshi1997@gmail.com

Pamela Jha (Ed.)

SPECTROPHOTOMETRY OF DNA AND RNA

Aim

To determine the concentration of DNA or RNA in a sample.

Principle

There are several ways to quantitate solutions of nucleic acids. If the solution is pure, one can use a spectrophotometer to measure the amount of ultraviolet radiation absorbed by the bases [1, 3].

Procedure

1. Use distilled water or 1X TE as a solvent to suspend the nucleic acids, and place each sample in a quartz cuvette.

2. Zero the spectrophotometer with a sample of solvent. For more accurate readings of the nucleic acid sample of interest, dilute the sample to give readings between 0.1 and 1.0.

For a 1 cm pathlength, the optical density at 260 nm (OD_{260}) equals 1.0 for the following solutions:

- a 50 µg/mL solution of dsDNA
- a 33 µg/mL solution of ssDNA
- a 20-30 µg/mL solution of oligonucleotide
- a 40 µg/mL solution of RNA

Troubleshooting

Contamination of nucleic acid solutions makes spectrophotometric quantitation inaccurate. Calculate the OD_{260}/OD_{280} ratio for an indication of nucleic acid purity. Pure DNA has an OD_{260}/OD_{280} ratio of ~1.8; pure RNA has an OD_{260}/OD_{280} ratio of ~2.0. Low ratios could be caused by protein or phenol contamination.

COLOURIMETRIC ASSAYS

Nucleic acids are isolated by homogenizing fresh samples, disrupting the cells, and precipitating the DNA and RNA with trichloroacetic acid and ethanol. To determine concentrations of nucleic acids a colourimetric assay is used.

Diphenylamine Reaction

Aim

To determine the concentration of DNA by the Diphenylamine (DPA) method.

Principle

Diphenylamine is specific for 2'-deoxypentose, including the deoxyribose characteristic of DNA. When a solution of diphenylamine and DNA is heated, the solution turns blue. The intensity of the blue colour, measured at 600 nm, is directly proportional to the concentration of the DNA. the lack of a hydroxyl substituent at the 2' position enables a ring-opening oxidation reaction, the aldehyde product of which further reacts with diphenylamine [$(Ph)_2NH$] to yield a bright blue complex (Fig. **12.1**) [1, 2].

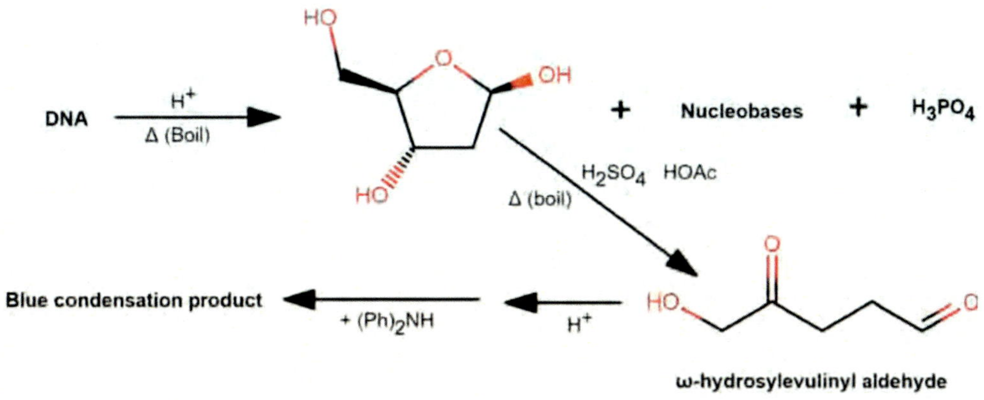

Fig. (12.1). In Dische's assay.

Procedure

Label six test tubes as shown in Table **12.1**, and add the appropriate volume of each solution. Tubes 1-4 are the standards, which contain known concentrations of DNA; tube 5 contains the extract prepared above with an unknown concentration of DNA. Make sure that you use the correct concentration of each stock solution. The problems, causes and probable solutions are mentioned in Table **12.2**.

Table 12.1. DPA method procedure.

Tube	The volume of DNA (µg/ml)	Stock Solution	Nucleic Acid Extract	5 % TCA
Blank	0	-	-	2.0 ml

(Table 12.1) cont.....

Tube	The volume of DNA (µg/ml)	Stock Solution	Nucleic Acid Extract	5 % TCA
1	100	2.0 ml (100 µg/ml)	-	-
2	200	2.0 ml (200 µg/ml)	-	-
3	300	2.0 ml (300 µg/ml)	-	-
4	400	2.0 ml (400 µg/ml)	-	-
5	?	-	2.0 ml	-

1. To each tube, add 4.0 ml diphenylamine reagent, dispensed from a buret. Cover each tube with parafilm and invert two or three times to mix. Precaution: This reagent contains concentrated acetic acid and sulfuric acid. If it comes in contact with your skin, wash it off immediately with water.

2. Incubate the tubes in a boiling water bath for 10min. Monitor your tubes carefully so that they do not spill.

3. Allow the content to cool down to room temperature.

4. Measure the absorbance at 600 nm.

6. Insert the reference Blank and adjust the controls so that the absorbance, *A*, reads zero.

7. Extrapolate the absorbance and express the concentration using a standard curve (Fig. **12.2**).

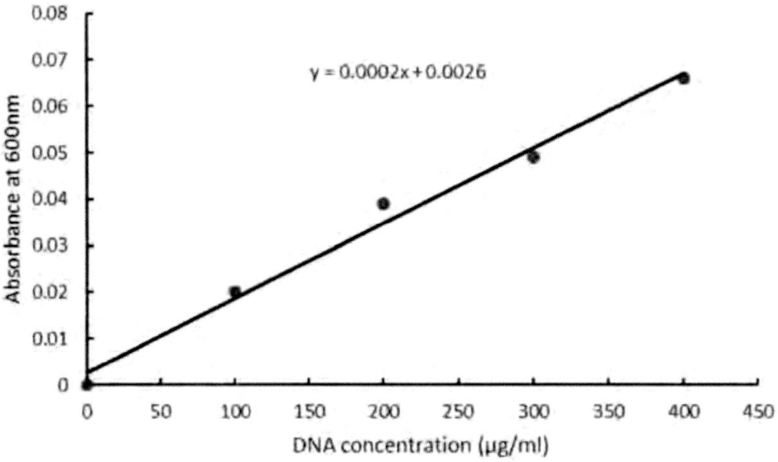

Fig. (12.2). Standard curve for DNA estimation by the DPA method.

Table 12.2. Problems, causes and solutions in DPA procedure.

Sr. no.	Problem	Possible Cause	Solution
1.	Standards and samples give lower OD values than expected, even though the Blank is correct	1. DPA reagent was not stored properly 2. The reagent was cold while performing the reaction 3. Absorbance was not measured at the correct wavelength	1. Always store reagent at 2-8°C 2. Allow the reagent to come to RT before starting the reaction 3. Measure the absorbance at the correct wavelength

Orcinol Reaction

Aim

To determine the concentration of RNA by the orcinol method.

Principle

Orcinol reacts with the ribose of RNA to produce a green-coloured solution. However, the orcinol reaction is not as specific as the diphenylamine reaction. Upon heating and acidification, sugars containing pentose rings will decompose to furfural, which then reacts with orcinol (Bial's reagent) to yield a soluble blue-green complex (Fig. **12.3**). To some extent, orcinol reacts with the deoxyribose of DNA as well, but the colour intensity of this reaction is only about 10% of that of the reaction with ribose of RNA. Therefore, it is necessary to subtract the contribution to the orcinol colour reaction made by DNA. The standard curve for the orcinol reaction is generated from absorbance readings taken at 660 nm [1] [3].

Procedure

1. Label six test tubes as shown in Table **12.3** and add the appropriate volume of each solution. Tubes 1-4 are the standards, which contain known concentrations of DNA; tube 5 contains the extract prepared above with an unknown concentration of RNA. Make sure that you use the correct concentration of each stock solution.

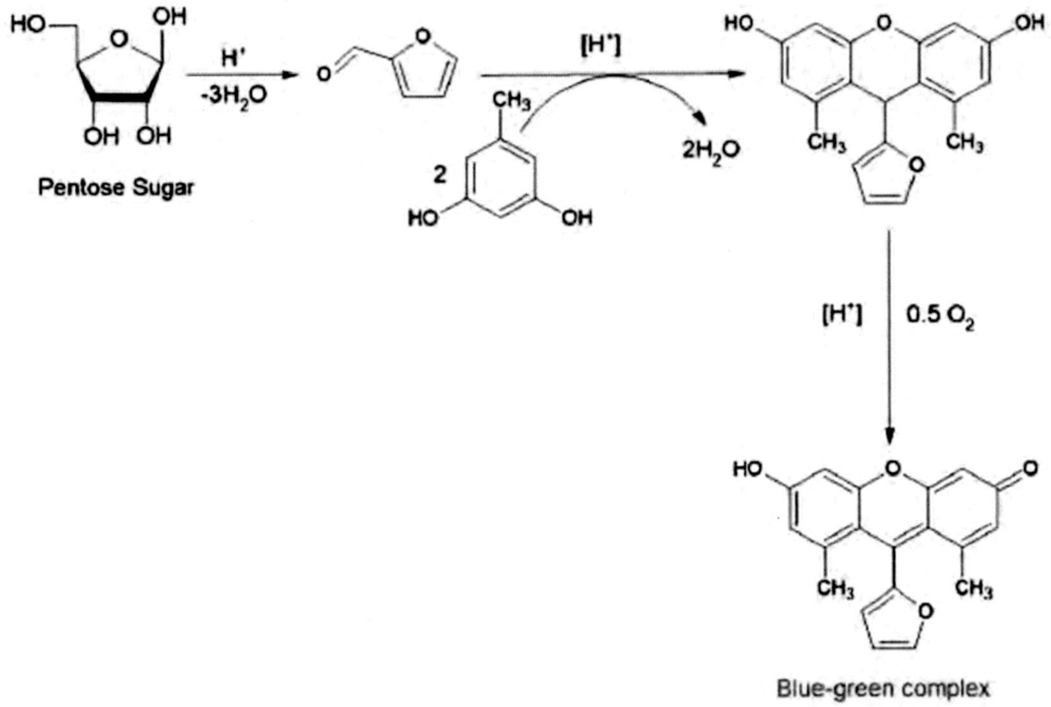

Fig. (12.3). Orcinol reaction.

Table 12.3. Orcinol procedure table.

Tube	[RNA] Volume of RNA (µg/ml)	Stock Solution	Nucleic Acid Extract	5 % TCA
Blank	0	-	-	3.0 ml
1	100	0.4 ml (100 µg/ml)	-	2.6 ml
2	200	0.4 ml (200 µg/ml)	-	2.6 ml
3	300	0.4 ml (300 µg/ml)	-	2.6 ml
4	400	0.4 ml (400 µg/ml)	-	2.6 ml
5	?	-	0.4 ml	2.6 ml

2. To each tube, add 3.0 ml of the orcinol reagent, dispensed from a buret. Cover each tube with parafilm and invert two or three times to mix.

3. Place the tubes in a boiling water bath and incubate for 20 minutes. Monitor your tubes carefully so that they do not spill.

4. After incubation, allow tubes to cool down at room temperature, and measure the OD at 660nm.

5. Extrapolate the absorbance and express the concentration using a standard curve (Fig. **12.4**). The problems, causes and probable solutions are mentioned in Table **12.4**.

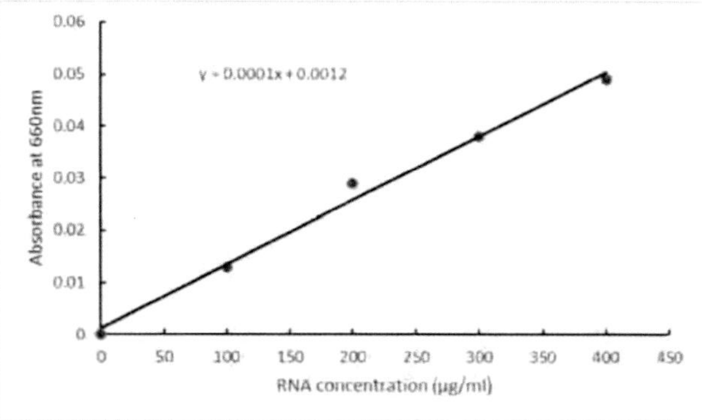

Fig. (12.4). Standard curve for RNA estimation by the Orcinol method.

Precautions

1. Orcinol reagent contains concentrated hydrochloric acid. If it comes in contact with your skin, wash it off immediately with water.

2. The assay should be carried out at the same time and in the same buffer conditions.

3. The Diphenylamine reagent can be prepared in advance. Once prepared, it should be stored at 2-8° C.

4. The orcinol reagent should be prepared freshly and stored at RT in an amber-coloured bottle.

Table 12.4. Problems, causes and solutions in Orcinol reaction.

Sr. no.	Problem	Possible Cause	Solution
1.	Standards and samples give lower OD values than expected although the Blank is ok	1. Orcinol reagent was not prepared freshly 2. Absorbance was not measured at the correct wavelength	1. Prepare the Orcinol reagent freshly, before use 2. Measure the absorbance at the correct wavelength

Did you know?

1. If you could type 60 words per minute, eight hours a day, it would take

approximately 50 years to type the human genome.

2. DNA is a fragile molecule. About a thousand times a day, something happens to it to cause errors. This could include errors during transcription, damage from ultraviolet light, or any of a host of other activities. There are many repair mechanisms, but some damage isn't repaired. This means you carry mutations! Some of the mutations cause no harm, a few are helpful, while others can cause diseases, such as cancer.

3. Friedrich Miescher discovered DNA in 1869, although scientists did not understand DNA was the genetic material in cells until 1943. Before that time, it was widely believed that proteins stored genetic information.

QUANTIFICATION OF NUCLEIC ACIDS BY ADVANCED FLUORESCENCE TECHNIQUES

There are 2 major fluorophores used for the estimation of proteins by advanced fluorescence techniques: Epicocconone and Nanoorange [5, 6].

DNA Detection using the DNA-binding Fluorochrome Hoechst 33258:

Non-intercalating dyes, Hoechst, bind to the minor groove of DNA in A–T-rich regions. These dyes effectively stain the DNA of living and fixed cells and can be applied to immunohistochemistry. The lambda max of Hoechst 33258 is ~461 nm. The Hoechst 33258 fluorochrome works equally well with whole cells and purified DNA and is specific for nanogram amounts of DNA. The fluorescence also has little affinity for RNA.

Principle

Fluorophores or other small molecules of interest are fused to Hoechst via a short flexible linker. The high affinity of the Hoechst moiety for DNA allows these composite probes to enter the cell and accumulate inside its nucleus. Hoechst 33258 binds to DNA in two different modes: the high affinity (Kd 1–10 nM) binding results from the specific interaction with B-DNA minor groove and the low affinity (Kd ~1000 nM) reflects the nonspecific interaction with DNA sugar-phosphate backbone (Fig. **12.5**). The -OH group on the linker binds to the moiety using standard Williamson ether synthesis conditions. It can also be readily reduced into triflate and then coupled to a moiety by a transition metal-mediated cross-coupling reaction. The fluorochrome is, however, sensitive to changes in DNA composition, with preferential binding to AT-rich regions [4 - 6]. A fluorometer is capable of an excitation wavelength of 365 nm and an emission wavelength of 460 nm is required for this assay.

Range of DNA detected: 10 ng/ml to 15 μg/ml

Advantages

• Hoechst dyes are among the most popular fluorophores used to stain DNA in living and fixed cells.

• The use of fluorometry to measure DNA levels has become popular primarily due to its simplicity and sensitivity compared to spectrophotometric measurements.

• Their high affinity and specificity toward DNA make Hoechst dyes excellent targeting moieties, which can be conjugated to various other molecules to tether them to DNA.

• Low cost and simplicity of use.

• Concentrations of DNA in both crude cell lysates and purified preparations can be quantified (Labarca and Paigen, 1980).

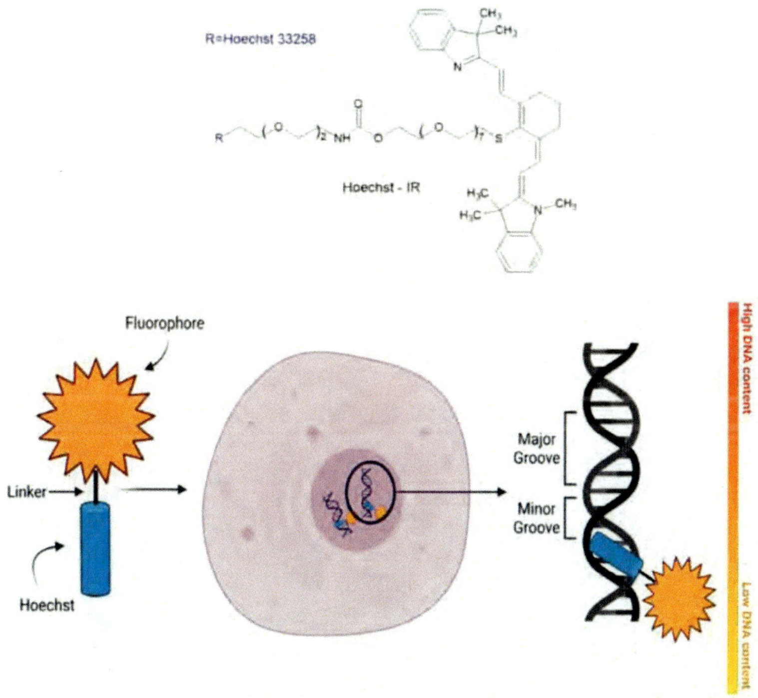

Fig. (12.5). DNA detection by fluorochrome.

Disadvantages

- Imaging with Hoechst in the UV/blue region has serious disadvantages of phototoxicity of UV light.
- This hampers Hoechst applications in living cells, tissues and animals and may pose a major problem when detecting relatively small amounts of non-nuclear DNA.
- The assay is affected by differences in the base composition of DNA.

DNA And RNA Detection with Ethidium Bromide Fluorescence

Ethidium bromide is not as sensitive as Hoechst 33258 and, although capable of detecting nanogram levels of DNA, will also bind to RNA.

Principle

Ethidium bromide is a DNA intercalator that inserts itself between base pairs in the double helix. At 300 and 360 nm, ethidium bromide has maximum UV absorbance, while its maximum emission is at 590 nm. The staining of ssDNA or RNA is less sensitive and requires 10 times more nucleic acid for a similar result. As a result of the hydrophobic environment between the DNA base pairs, ethidium bromide displays intense fluorescence when bound to DNA. By moving into this environment and away from the solvent, the EtBr cation (Fig. **12.6**) sheds any water molecules attached to it. The removal of water molecules allows EtBr to fluoresce since water is a highly efficient fluorescent quencher [4 - 6]. It is also possible to detect ssRNA because it folds back onto itself and therefore provides local base pairing for dye intercalation. A fluorometer is capable of an excitation wavelength of 302 or 546 nm and an emission wavelength of 590 nm is required for this assay.

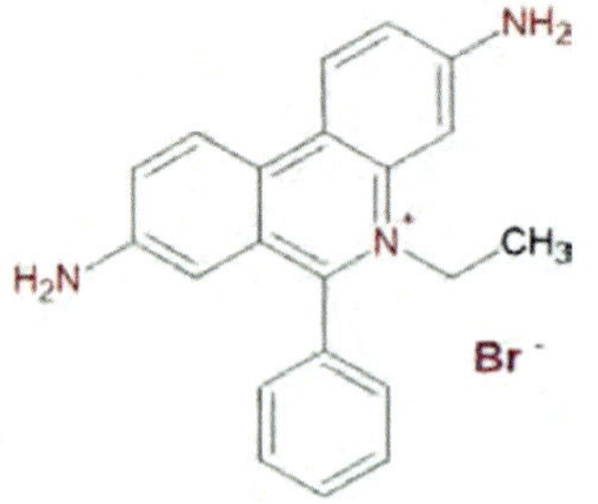

Fig. (12.6). Structure of EtBr.

Range of detection for DNA: 0.5 to 5.0 ng/band.

Advantages

• Ethidium bromide is regularly used to stain DNA on gels, but it has also been used to detect protein-DNA complexes in band shift assays, as well as to observe single DNA molecules in gel electrophoresis.

• In contrast to the fluorochrome Hoechst 33258, ethidium bromide is relatively unaffected by differences in the base composition of DNA.

• In preparations of DNA with minimal RNA contamination or with DNA samples having an unusually high guanine and cytosine (GC) content, ethidium bromide offers a relatively sensitive alternative to the more popular Hoechst 33258 DNA assay.

Disadvantages

• EtBr may act as a strong mutagen. Following its toxicity, it is defined as harmful by all routes of entry; inhalation, ingestion, or skin absorption.

Did you know?

• Phosphorescence is the emission of light from a substance exposed to radiation and persisting as an afterglow after the exciting radiation has been removed. It is a characteristic feature of phosphorescence that, unlike fluorescent light, which is spontaneously emitted about 10-8 seconds following excitation, phosphorescence is induced by additional excitation and can last anywhere between 10-3 seconds and days or years, depending on the circumstances.

• In the early 1900s, the first uses of fluorophores in biological investigations were performed to stain tissues, bacteria, and other pathogens. This was later developed into fluorescence microscopy by Carl Zeiss and Carl Reichert.

DNA Detection using DNA-binding Fluorochrome PicoGreen

Principle

PicoGreen is a fluorometric method that is a highly sensitive means of dsDNA detection and quantitation with minimum consumption of samples. PicoGreen selectively binds dsDNA and exhibits maximum emission of fluorescence at 530nm whereas, when unbound it exhibits minimal fluorescence [7 - 9]. Due to this reason, it is considered more sensitive than other dyes and can detect nucleic acid (specifically dsDNA) concentrations as low as 25ng/ml. Therefore, PicoGreen provides highly sensitive quantitation by consuming minimal samples [7 - 9].

Requirements

Equipment- Flurospectrometer, Micropipette

Chemicals- 1X TE Buffer, PicoGreen dye stock, standard dsDNA samples, unknown dsDNA samples.

Glassware- Nuclease-free amber or foil covered 1.5 ml polypropylene tubes

Others: Pipette tips, Nuclease-free distilled water (D/W)

Procedure

1. Dilute the PicoGreen concentrated stock 1:200 by transferring 10 μl of dye stock and 2 ml of 1X TE to an amber tube and prepare the PicoGreen working solution.

2. Mix well and protect from light.

3. Thaw the dsDNA known/standard and unknown samples and mix each solution properly but gently.

4. Serially dilute dsDNA standards to either 2μg/ml (1:50) or 50ng/ml (1:1000) in nuclease-free tubes.

5. Add 2 μl of unknown sample in 100 μl of TE buffer in nuclease-free tubes to determine the amount of sample for the assay.

6. In labelled nuclease-free amber or foil-covered tubes, add an equal volume of the diluted standard samples and unknown samples.

7. Add an equal volume (100μl) of the PicoGreen working solution to each tube containing samples and standards.

8. Mix all the prepared tubes thoroughly and allow them to rest at room temperature for 5 minutes in a dark place.

9. The samples and standard tubes were read using fluorospectrometer, at 485nm excitation and 535nm emission filters.

Results

Example of the spectrum and the standard curve obtained from a PicoGreen sample (Fig. **12.7**).

Q-PCR

Principle

q-PCR is a powerful technique used for both qualitative and quantitative analysis of nucleic acids due to its high sensitivity and specificity. The main principle of any PCR, including q-PCR, is similar to traditional PCR that uses the principle of enzymatic replication of nucleic acids [9, 10]. Here, the first denaturation of dsDNA takes place, followed by the annealing of primers to the ssDNA template strand and extension of the complementary strand formed, thus, amplifying the target DNA during the thermal cycling process [9, 10]. The unique difference of q-PCR is due to the fluorometric probes like Taqman and DNA binding dyes like SYBR® Green, which are used to bind target DNA during the annealing phase of PCR [9, 10]. These probes are cleaved and displaced, which results in the emission of fluorescent dye upon extension of the target DNA sequence. The emission of fluorescence of these dyes detected to determine the success of the amplification reaction and the initial quantity of nucleic acids present in the sample, which is represented in a graphical form.

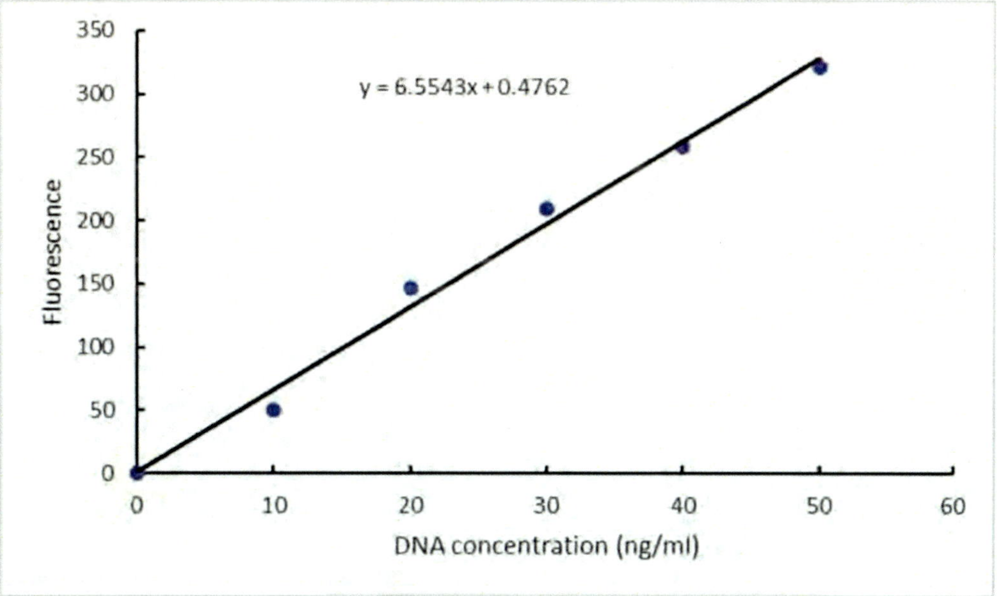

Fig. (12.7). The standard curve of PicoGreen standards.

Requirements

Equipment- Thermal cycler for q-PCR, Micropipettes, Vortex, Spin Down centrifuge

Chemicals- 10X q-PCR buffer, 50mM $MgCl_2$, 20mM dNTPs, 10μM Primers (Forward and Reverse), SYBR Green I dye, 5μM Taqman probe, 5 U/μl Taq Polymerase

Glassware- PCR tubes, PCR tube rack

Others: Nuclease-free distilled water (D/W)

Procedure

1. Prepare MasterMix as given in the Table **12.5** below:

Table 12.5. Ingredients with their concentrations in MasterMix.

MasterMix	Final Concentration	Volume Per Reaction (μl)	
-	-	SYBR Green I Dye	Taqman
dH_2O	-	14.125	12.125
10X buffer	1X	2.5	2.5
50mM $MgCl_2$	2.5mM (5mM for Taqman)	1.25	2.5
20mM dNTPs	0.8mM	1	1
10μM Primers (Forward & Reverse)	0.4μM	0.25	0.25
SYBR Green, I diluted 1:50,000	-	0.25	-
5μM probe	200nM	-	1
5 U/μl Taq Pol	-	0.25	0.25

Mix all the reagents gently by pipetting and then add Taq Pol (Polymerase) and mix gently again.

2. Add 5 μl of sample to the PCR tubes.

3. Add 20 μl of the master mix prepared to all the tubes with the sample. Vortex and short spin to ensure that no bubbles are formed.

4. Place the tubes in the q-PCR thermal cycler.

5. For SYBR Green I dye, set and perform the 3-step PCR reaction according to the Table **12.6** given below:

6. Analyze and interpret the graph obtained.

Table 12.6. Steps with specifications for PCR Run.

Cycles	Steps	Temperature	Time
1 cycle	Activation	95°C	10min
-	Denaturation	95°C	15s
40 cycles	Annealing	50-65°C	30s
-	Extension	72°C	30s

For labelled probes, set and perform the 2-step PCR reaction according to the Table **12.7** given below:

Table 12.7. Set up with specifications for PCE run with labelled probes.

Cycles	Steps	Temperature	Time
1 cycle	Activation	95°C	10min
40 cycles	Denaturation	95°C	15s
-	Annealing/Extension	50-65°C	1min

Results

Example of amplification plot of standard curves of the q-PCR method.

Advantage

• Ability to assess the presence and amount of the target DNA present in the sample.

• Ability to determine the concentration of nucleic acids by comparing the amplification plot of the sample with the standard curve of known concentrations.

• It detects PCR inhibitors within the reaction mixture and has high specificity due to the use of fluorometric probes.

Disadvantages

• This assay uses expensive reagents, primers, and probes, and the thermal cycler of q-PCR is also expensive.

• The time required for the reaction to take place is lengthy.

• Measurement accuracy may depend on the assay design.

CONCLUSION

The quantitative analysis of nucleic acids has been discussed in this chapter. These are some of the most common and standard procedures for quantification. Some advanced techniques using fluorescence are also discussed. Depending on the availability and requirements, any one method can be followed.

VIVA VOCE

1. How do you determine the purity of an isolated nucleic acid?

Ans. To evaluate DNA purity, measure absorbance from 230 nm to 320 nm to detect other possible contaminants. The most common purity calculation is the ratio of the absorbance at 260 nm divided by the reading at 280 nm. Good-quality DNA will have an A_{260}/A_{280} ratio of 1.7–2.0.

2. What does a 260/280 ratio mean?

Ans. Abnormal 260/280 ratios usually indicate that a sample is contaminated by residual phenol, guanidine, or other reagent used in the extraction protocol, in which the ratio is normally low. Inaccurate ratios may also be encountered at very low concentrations (< 10 ng/µl) of nucleic acids.

3. What is the chemical reaction involved in the diphenylamine test?

Ans. DNA can be identified chemically with the Dische diphenylamine test. Acidic conditions convert deoxyribose to a molecule that binds with diphenylamine to form a blue complex. The intensity of the blue colour is proportional to the concentration of DNA.

4. Why is orcinol used in RNA estimation?

Ans. One of the widely used reactions is those of RNA with Orcinol. So, the basic objective of this test is to estimate RNA by the orcinol method. The method usually involves the pentose conversion and ribose in the presence of hot acid to furfural that again reacts with orcinol to form a green colour.

5. What are the applications of the Hoechst 33258 fluorescence quantitation technique?

Ans. Hoechst dyes stain the DNA of both living and fixed cells and are compatible with immunohistochemistry applications.

6. What are the critical parameters to maintain the sensitivity of the Hoechst 33258 fluorescence assay?

Ans. The sensitivity of the Hoechst 33258 fluorescence assay decreases with nuclease degradation, increasing GC content, or denaturation of DNA (Labarca and Paigen, 1980; Stout and Becker, 1982). Increased temperature of the assay solution and ethidium bromide contamination also decrease the Hoechst 33258 signal. Sodium dodecyl sulfate (>0.01% final concentration) also interferes with accurate readings (Cesarone *et al.*, 1979). The pH of the assay solution is critical to sensitivity and should be pH ~7.4 (Stout and Becker, 1982; Labarca and Paigen, 1980).

7. How does pH affect the Hoechst 33258 fluorescence assay?

Ans. The assay solution's pH should be pH 7.4 to ensure sensitivity (Stout and Becker, 1982; Labarca and Paigen, 1980). At a pH <6.0 or >8.0, the background becomes much higher, and there is a concomitant loss of fluorescence enhancement.

8. How do different forms of DNA/RNA contribute to the EtBr fluorescence assay?

Ans. In the ethidium bromide assay, single-stranded DNA gives approximately half the signal as double-stranded calf thymus DNA. Ribosomal RNA also gives about half the fluorescent signal as double-stranded DNA, and RNase and DNase both severely decrease the signal. Closed circular DNA also binds less ethidium bromide than nicked or linear DNA.

9. Why PicoGreen is considered to be a sensitive dye for DNA detection?

Ans. PicoGreen selectively binds to the dsDNA and emits maximum fluorescence at 530 nm and minimum when undone.

10. What is the detection limit of the PicoGreen method?

Ans. It can detect dsDNA as low as 25ng/ml.

11. What is the difference between PCR and qPCR?

Ans. The q-PCR utilizes fluorometric probes like Taqman and SYBR® Green, which bind to the target DNA during the annealing. When these probes are cleaved and displaced, it results in the emission of fluorescent dye upon extension of the target DNA sequence.

REFERENCES

[1] Patterson, J.; Mura, C. Rapid colorimetric assays to qualitatively distinguish RNA and DNA in biomolecular samples. *J. Vis. Exp.,* **2013**, (72), e50225.

[PMID: 23407542]

[2] Gendimenico, G.J.; Bouquin, P.L.; Tramposch, K.M. Diphenylamine-colorimetric method for DNA assay: A shortened procedure by incubating samples at 50°C. *Anal. Biochem.,* **1988**, *173*(1), 45-48. [http://dx.doi.org/10.1016/0003-2697(88)90156-X] [PMID: 3189801]

[3] De Mey, M.; Lequeux, G.; Maertens, J.; De Maeseneire, S.; Soetaert, W.; Vandamme, E. Comparison of DNA and RNA quantification methods suitable for parameter estimation in metabolic modeling of microorganisms. *Anal. Biochem.,* **2006**, *353*(2), 198-203. [http://dx.doi.org/10.1016/j.ab.2006.02.014] [PMID: 16545766]

[4] Mackintosh, J.A.; Veal, D.A.; Karuso, P. Fluoroprofile, a fluorescence-based assay for rapid and sensitive quantitation of proteins in solution. *Proteomics,* **2005**, *5*(18), 4673-4677. [http://dx.doi.org/10.1002/pmic.200500095] [PMID: 16267819]

[5] Silberblatt, S.; Felder, R.A.; Mifflin, T.E. Optimizing reaction conditions of the NanoOrange® protein quantitation method for use with microplate-based automation. JALA. *J. Assoc. Lab. Autom.,* **2001**, *6*(6), 83-88. [http://dx.doi.org/10.1016/S1535-5535-04-00169-8]

[6] Gallagher, S.R. Quantitation of DNA and RNA with absorption and fluorescence spectroscopy. *Curr. Protoc. Mol. Biol.,* **2011**, *3*(Appendix), 3D. [PMID: 21225635]

[7] Caporaso, J. G.; Ackermann, G.; Apprill, A.; Bauer, M.; Berg-Lyons, D.; Betley, J.; Weber, L. EMP 16S Illumina amplicon protocol. **2018**. Available From: https://www.who.int/news-room/fac--sheets/detail/monkeypox

[8] Dragan, A.I.; Casas-Finet, J.R.; Bishop, E.S.; Strouse, R.J.; Schenerman, M.A.; Geddes, C.D. Characterization of picogreen interaction with dsDNA and the origin of its fluorescence enhancement upon binding. *Biophys. J.,* **2010**, *99*(9), 3010-3019. [http://dx.doi.org/10.1016/j.bpj.2010.09.012] [PMID: 21044599]

[9] Shen, C. H. Detection and analysis of nucleic acids. *Diagn. Mol. Pathol.,* **2019**, 167-185.

[10] Nolan, T.; Hands, R.E.; Bustin, S.A. Quantification of mRNA using real-time RT-PCR. *Nat. Protoc.,* **2006**, *1*(3), 1559-1582. [http://dx.doi.org/10.1038/nprot.2006.236] [PMID: 17406449]

CHAPTER 13

Purification of Nucleic Acids

Sai Joshi[1,*], **Sinchan Hait**[2] and **Niriksha Patel**[3]

[1] *CSIR-NCL, Dr. Homi Bhabha Road, Pune - 411 008, India*

[2] *Department of Chemical Engineering, BITS-Pilani K.K. Birla Goa Campus, Goa-403726, India*

[3] *Symbiosis Internationlq University, Pune, India*

Abstract: Besides qualitative and quantification of nucleic acids, purification is an integral part of any genetic engineering experiments. In genetic engineering experiments, the purity of nucleic acid matters the most. Nucleic acid purification can be achieved by the phenol-chloroform method, alcohol-based purification, agarose gel electrophoresis, and chromatographic purification. These purifications are based on the interactions between bases of nucleic acids and chemical reagents. In this chapter, nucleic acid purification is explained in detail.

Keywords: Alcohol, Agarose gel electrophoresis, Chromatography, Genetic engineering, Nucleic acid, Phenol-chloroform.

INTRODUCTION

Five basic steps of DNA extraction are consistent across all the possible DNA purification chemistries [1]:

1. Disruption of the cellular structure to create a lysate.

2. Separation of the soluble DNA from cell debris and other insoluble material.

3. Binding the DNA of interest to a purification matrix.

4. Washing proteins and other contaminants away from the matrix.

5. Elution of the DNA.

* **Corresponding author Sai Joshi:** CSIR-NCL, Dr. Homi Bhabha Road, Pune - 411 008, India; E-mail: saijoshi1997@gmail.com

Pamela Jha (Ed.)

SEPARATION OF SOLUBLE DNA FROM CELL DEBRIS AND OTHER INSOLUBLE MATERIAL

Extraction of the Nucleic Acid Sample by Phenol-chloroform Method

Principle

DNA is insoluble in phenol because phenol is a nonpolar solution having a density of 1.07 g/cm^3, which is higher than the density of water (1.00 g/cm^3), thus forming two separate phases - an organic and aqueous phase when added to the cell sample. Chloroform protects genomic DNA and increases the efficiency of phenol to denature the protein. It allows proper separation of the organic phase and aqueous phase and keeps DNA protected in the aqueous phase. The mixing of phenol and chloroform forms a denser solution, thus enabling less cross-contamination from the organic phase in the aqueous phase, which is beneficial when the aqueous phase is removed from the solution to obtain a pure nucleic acid sample [2].

Binding to the Purification Matrix

The DNA of interest can be isolated using various methods and its purification can be performed by binding to the matrices using silica, cellulose and ion exchange depending on their characteristic binding capacity and efficiency.

Solution-Based Chemistry

It relies on alcohol precipitation rather than relying on a binding matrix. The cell debris and proteins are precipitated using a salt solution of high concentration which will remove proteins from the solution and centrifugation would separate the soluble nucleic acid from the cell debris and precipitated protein. The DNA is precipitated using isopropanol. Removal of remaining salt can be done by washing the pellet with ethanol and the DNA pellet is resuspended in an aqueous buffer like Tris-EDTA and is then ready for its use in downstream applications.

ALCOHOL-BASED PURIFICATION METHOD

Aim

To purify nucleic acid samples using alcohol.

Principle

The reaction between solute and solvent creates an insoluble aggregate called precipitate and the process is called precipitation.

DNA is a polar molecule with a net negative charge on the backbone (PO_3^-). In DNA precipitation, salt (sodium acetate) reacts with DNA and breaks up into Na^+ and CH_3COO^-. The positively charged sodium ion neutralizes negatively charged PO_3^- of the DNA. The hydrophilic nature of DNA helps it to dissolve in the water, but reacting with sodium acetate makes DNA less hydrophilic. According to Coulomb's law, "force of attraction between two opposite charges is inversely proportional to the dielectric constant." The low dielectric constant of ethanol protects the salt - DNA complex by shielding it against the water. Therefore, the chance of DNA getting dissolved in alcohol is less as compared to that of water and after adding alcohol, the force of interaction between PO_3^- and Na^+ increases. Alcohol forms hydrogen bonding with water, and thus, it protects the complex of PO_3^- and Na^+ by neutralizing the charge of water. Collectively, the DNA is pulled out from the bottom of the tube as a white cottony thread-like precipitate [3].

The most common salts utilized in DNA precipitation are sodium acetate, sodium chloride, ammonium acetate, and lithium chloride.

Lithium chloride is best used for RNA precipitation, whereas sodium acetate is employed for DNA precipitation.

Requirements

Glassware: test tubes, microcentrifuge tube, beaker.

Reagents: DNA sample (<1mg/ml), 3 M sodium acetate (pH 5.2), 1:1 phenol-chloroform, ice-cold 100% ethanol, 70% ethanol, TE buffer (pH 8.0).

Other Requirements: vortex, microcentrifuge, speed vac evaporator, pipettor.

Procedure

1. Take the DNA solution to be purified and add an equal volume of phenol and chloroform and place it in a 1.5ml microcentrifuge tube.

2. At room temperature, vortex vigorously for 10 seconds and microcentrifuge for 15 seconds at maximum speed so that the organic phase and aqueous phase are well distinct.

3. Carefully remove the aqueous phase containing DNA from the top using a 200μl pipettor and transfer it to a new tube. If a white precipitate is observed at the aqueous/organic interface, repeat steps 1 to 3.

4. Add 1/10 volume of 3 M sodium acetate of pH 5.2 to the solution of DNA. Mix by vortexing briefly or by flicking the tube several times with a finger.

5. Add 2 to 2.5 volumes of ice-cold 100% ethanol. Vortex the solution and place it in a −70°C freezer for 15-20 minutes.

6. Centrifuge for 5 minutes at maximum speed and remove the supernatant carefully.

7. Add 1 ml of 70% ethanol at room temperature. Invert the tube several times and microcentrifuge again for 5 minutes at maximum speed.

8. Remove the supernatant and dry the pellet in the Speedvac evaporator.

9. Dissolve the dry pellet in an appropriate volume of water if it is going to be used for further enzymatic manipulations requiring specific buffers, whereas dissolving in the TE buffer of pH 8.0 is going to be stored indefinitely.

Precaution

Leave the washed pellet on the lab table carefully for a few minutes to avoid over-drying the sample, which will further denature the DNA.

If isopropanol is used instead of ethanol during extraction, then the sample may not adhere tightly to the tube and would require a longer time to dry.

Did you know?

Kiwis, strawberries and bananas work very well to extract DNA.

Silica-Binding Chemistry

It is based on the binding of DNA to silica under high salt conditions in the presence of chaotropic salt like guanidine hydrochloride. The nucleic acid is washed with salt to remove undesired components and the bound DNA can then be eluted under low salt conditions either by nuclease-free water or TE buffer.

Pure DNA can further be obtained by adding ribonuclease RNase A to the elution buffer, which would copurify with genomic DNA. The addition of RNase would ensure the removal of RNA, leading to the retrieval of pure DNA.

SILICA COLUMN-BASED PURIFICATION

Aim

To purify the nucleic acid sample using the silica column method.

Principle

The spin-column tube is formed from three major parts: the silica as a solid phase that is settled inside the column tube, the column and the collection tube. It is based on the principle of affinity between positively charged silica material and negatively charged DNA leading to the selective binding of nucleic acids to the silica matrices, while the rest of the cell components, chemicals and impurities are washed out from the column. The chemistry of the spin column method relies on the pH and salt concentration. Once the DNA is bound to silica in the presence of desired pH and high concentration of chaotropic salts like guanidine HCL, guanidine thiocyanate, urea, and lithium perchlorate, it cannot pass through the column and other impurities get collected into the collection tube [4, 5].

Elution of DNA from the silica matrix is done by employing a hypoosmotic solution such as alkaline buffer Tris-EDTA, and thus, DNA gets dissolved into the buffer. The dissolved DNA detached from the silica is collected into the collection tube by the process of centrifugation.

The major advantage of silica membrane spin columns is that the silica is bound to a solid support which eradicates the problem of glass-bead contamination of the DNA sample, whereas the downside of this technique is being unable to efficiently recover small DNA fragments as they are tightly bound with the silica matrix [4, 5].

Sodium plays a role as a cation bridge that attracts the negatively charged oxygen in the phosphate backbone of nucleic acid. Sodium cations break the hydrogen bonds between the hydrogen in water and the negatively charged oxygen ions in silica under high salt conditions (pH $\leq$7). the role played by sodium ions in attracting the negatively charged oxygen present in nucleic acid's phosphate group and helping NA become insoluble because of the phenomenon known as “salting out” in the presence of high-salt conditions and acidic pH.

Chaotropes have two major roles:

1. They destabilize hydrogen bonds, van der Waal forces and hydrophobic interactions resulting in the destabilization of nucleases and proteins.

2. The association of nucleic acids with water is disrupted, thereby providing optimal conditions for their transfer to silica.

Requirements

Glasswares

Silica membrane spin columns, 1.5-ml microcentrifuge tubes.

Other Requirements

DNA sample, microcentrifuge, vortex.

Reagents

Resuspension buffer, Neutralization/binding solution, Wash buffer, TE buffer (pH 8.5).

Preparation of Reagents

1. Resuspension buffer - 50 mM Tris HCl (pH 8.0), 10 mM EDTA, 100 mg RNase A. Store indefinitely at 4°C.

2. Neutralisation/binding solution - Add 477.65 g guanidine HCl and 49.09 g potassium acetate to 500 ml distilled water and stir to dissolve. Adjust pH to ~4.2 with the acetic acid, dilute solution to 1 litre with water, and filter sterilize. Store indefinitely at 4°C.

3. Wash buffer - 80% ethanol, 20 mM NaCl, 2 mM Tris HCl (pH 7.5). Store indefinitely at room temperature.

4. Lysis solution - 0.2 M NaOH/1.0% (w/v)

Procedure

Preparation of DNA Solution

1. Harvest 1 to 5 ml of bacterial culture at room temperature by centrifuging for 1 min at 10,000 × g, then discard the supernatant.

2. Resuspend the cell pellet by adding 250 µl of cell resuspension buffer and vortexing. further, transfer it to a 1.5 ml microcentrifuge tube.

3. Add 250 µl lysis solution and mix by inversion. Allow lysis to proceed for 3 to 5 minutes.

4. Then add 350 µl neutralization/binding solution and blend by inversion.

5. Microcentrifuge for 10 minutes at maximum speed. A white pellet will be visible at the bottom of the tube.

Purifying DNA on Silica Membrane Spin Column

6. Pour the supernatant into the silica membrane spin column.

7. Microcentrifuge spin column inside its collection tube for 1 minute at maximum speed. Then remove the spin column from the collection tube and discard its flowthrough. Reinsert the spin column into the collection tube.

8. Now add a 750 µl wash buffer to wash the spin column and microcentrifuge for 1 minute at maximum speed. Remove the spin column from the collection tube again and discard the flowthrough.

9. Reinsert the spin column within the collection tube and microcentrifuge for an additional 1 minute to get rid of any residual wash buffer from the column membrane.

10. Transfer the spin column to a 1.5 ml microcentrifuge tube and add 75 to 100 µl TE buffer of pH 8.5 to the center of the membrane. Wait for 2 to 10 minutes and then microcentrifuge for 1 minute at maximum speed.

11. Collect DNA and store at 4°C until use.

Results

DNA purification using a silica column is successfully done.

Precautions

1. Handle guanidine HCl with care; it may cause skin irritation and do not inhale.

Did you know?

The purified DNA is readily available for secondary PCR, restriction digests, sequencing, subcloning, ligations and other enzymatic manipulations.

It is used for the purification of genomic DNA from animal, plant and microbial sources and also for the purification of RNA, plasmids and PCR products.

Cellulose-Binding Chemistry

Nucleic acids bind to cellulose in the presence of high salt and alcohol. Conditions can be adjusted to preferentially bind different sizes and species of nucleic acid to the cellulose to obtain high concentration eluates for nucleic acid.

AGAROSE GEL ELECTROPHORESIS BASED PURIFICATION

Aim

To purify nucleic acid samples using agarose gel electrophoresis.

Principle

Agarose gel electrophoresis is one of the most effective ways to the separation of different sizes of DNA fragments ranging from 100 bp to 25 kb. The DNA is loaded into precast wells within the gel, and a current is applied to separate them. DNA (and RNA), being negatively charged molecules due to the presence of phosphate backbone, will migrate from cathode to anode (positively charged) when placed in an electric field. The unique property of DNA of having a uniform mass/charge ratio enables it to be separated by size within the agarose gel in such a way that distance travelled is inversely proportional to the log of its molecular weight and therefore, DNA molecules of different sizes are separated into distinct bands during electrophoresis. The gel matrix in electrophoresis acts as a sieve: smaller DNA molecules will migrate faster than larger ones. After separation, the DNA molecules are often visualized under UV light after staining with a suitable dye (ethidium bromide) [6].

The DNA moves through the gel at different rates depending on its conformation. Supercoiled plasmid DNA having compact conformation enables it to move faster than a linear DNA fragment of the same size with the open circular form.

Requirements

Glasswares

Erlenmeyer flask.

Other Requirements

Bunsen flame, power supply, gel box, paper towels, UV light.

Reagents

Nucleic acid sample, agarose, TAE/TBE, ethidium bromide, loading dye, DNA size marker.

Preparation of Reagents

<u>Running Buffer</u>

TAE - 40 mM Tris-acetate, 1 mM EDTA

TBE - 45 mM Tris-borate, 1 mM EDTA

Loading dye- 0.25% bromophenol blue, 0.25% xylene cyanol, 30% glycerol

Procedure

Preparation of Gel

a. Weigh out the appropriate mass of agarose into an Erlenmeyer flask. Add a running buffer (Tris-acetate EDTA/Tris-borate EDTA) to the flask containing agarose and swirl to mix. The volume of the buffer should not be greater than 1/3 of the capacity of the flask.

b. Dissolve the agarose/buffer mixture completely by employing a Bunsen flame and remove the flask every 30 seconds to swirl the contents. Then add ethidium bromide (EtBr) to a concentration of 0.5 μg/ml.

c. Allow the buffer mixture to cool down either by incubating it in a 65 °C water bath or on the benchtop.

d. Place the gel tray into the casting apparatus. Alternatively, one might also tape the open edges of a gel tray to make a mould. Now place an appropriate comb to form wells into the gel mould. Pour the molten agarose and allow it to sit at room temperature. Remove the comb and carefully put the gel within the gel box.

Setting up of Gel Apparatus and Separation of DNA Fragments

a. Add loading dye of 6X concentration to the DNA samples to be separated.

b. Then add a running buffer enough to cover the gel surface. Attach the leads of the gel box to the power supply and switch it on to supply the desirable voltage (1-5V/cm between electrodes).

c. Remove the lid. Load the DNA sample(s) into the gel gradually and carefully.

An appropriate DNA size marker should even be loaded carefully together with experimental samples.

d. Replace the lid of the gel box. The cathode should be closer to the wells than the anode. Check if the electrodes are plugged into the correct slots in the power supply and switch them on.

e. Run the gel until the dye has migrated to an appropriate distance.

Visualizing Separated DNA Fragments

a. When electrophoresis has been completed, turn off the power supply and take away the lid of the gel box.

b. Also, remove the gel from the gel box and drain off the excess buffer from the gel surface. To eliminate extra running buffers, keep the gel tray on paper towels to absorb it.

c. Then remove the gel from the gel tray and expose it to UV light which is most typically done by employing a gel documentation system. DNA bands are visible as orange fluorescent bands.

Results

DNA purification using agarose gel electrophoresis is completed.

Precautions

As ethidium bromide is a suspected carcinogen, it should be properly disposed of as per institution regulations. Gloves must always be worn while handling gels containing EtBr.

It is better to position the gel on a glass plate rather than a plastic tray to safeguard the UV box, as the glass plate will not only reduce the UV significantly but also will protect the UV box from being cut by the razor blade.

Wearing proper UV protection is a must - especially for your eyes.

Did you know?

This is a commonly used technique for molecular cloning, like PCR- or restriction enzyme-based cloning.

Ion Exchange Chemistry

It is based on the interaction occurring between positively-charged particles and the negatively-charged phosphates in DNA under low salt conditions. The contaminants are washed using high-salt solutions while DNA is eluted under high-salt conditions.

Ion exchange chromatography is discussed in detail in section 5.5 of Chapter 5.

CONCLUSION

Several common and advanced techniques are discussed in this chapter. Depending upon the purification requirement, any one method can be chosen. Some of these techniques are costly as well.

VIVA VOCE

1. What is the function of 70% and 100% ethanol in DNA extraction?

Ans. The 70% ethanol allows the salts to dissolve while minimizing the DNA solubility. The function of 100% ethanol is to promote ethanol evaporation from DNA pellet, thus preventing any carryover.

2. Why are methanol and other alcohol not utilized in DNA precipitation?

Ans. The DNA precipitation mechanism is solely dependent on the dielectric constant of the solvent. The boiling point of methanol is less than that of ethanol and isopropanol, while the dielectric constant is higher. Since the dielectric point is inversely proportional to the precipitation efficiency, therefore methanol will not show good precipitation results.

3. Why is chilled ethanol better to use in DNA precipitation?

Ans. At lower temperatures, high "flok" or aggregates are obtained along with the addition of some of the precipitation agents. Therefore, the addition of chilled ethanol facilitates higher flocculation, thereby giving higher precipitation of DNA at a faster rate.

4. Why is it essential to add ethanol in the appropriate amount?

Ans. Adding ethanol in large quantities will bring down degraded material and small species, which will influence the absorbance at 260 nm (A260 readings), whereas too little ethanol may impede the washing of the salt from the membrane.

5. How is the lysis of plasmid isolation different from that of RNA or genomic DNA extraction?

Ans. The lysis for plasmid isolation is very different from RNA or genomic DNA extraction because plasmids must be separated from genomic DNA first. The addition of chaotropic will release all types of DNA at once, thereby losing the ability to differentiate between small circular DNA and high molecular weight chromosomes. Therefore chaotropic are not added until after cell lysis in plasmid isolation.

6. How do you get a better resolution of bands?

Ans. To increase the resolution (crispness) of DNA bands, the following parameters should be considered:

a) Running the gel for an extended period at a lower voltage

b) Usage of a wider gel comb

c) Loading less DNA within the well

7. How do you get a better separation of bands?

Ans. By adjusting the gel percentage, we can separate the similarly sized bands that are running too approximate to urge better separation. A higher percentage of agarose gel will enable the smaller bands to get separated, and at a lower percentage of gel, larger bands are often separated.

8. What factors affect the rate of migration of DNA through gel?

Ans. The rate of migration of a DNA molecule through the gel is set by the subsequent factors: 1) size of the DNA molecule; 2) concentration of agarose; 3) conformation of DNA; 4) voltage applied, 5) presence of ethidium bromide, 6) type of agarose and 7) electrophoresis buffer.

9. What is the utilization of loading dye in gel electrophoresis?

Ans. Loading dye adds density to the sample, thus allowing it to sink into the gel. It also provides colour thereby simplifying the loading process. They move at a standard rate through the gel, therefore, helping to estimate the distance travelled by the DNA fragments.

REFERENCES

[1] Sirakov, IN Nucleic acid isolation and downstream applications. *Nucleic Acids-From Basic Aspects to Laboratory Tools,* **2016**, 1-26.

[http://dx.doi.org/10.5772/61833]

[2] Sambrook, J; Russell, DW Purification of nucleic acids by extraction with phenol: Chloroform. *Cold Spring Harbor Protocols,* **2006**, *2006*(1).

[3] Li, Y.; Chen, S.; Liu, N.; Ma, L.; Wang, T.; Veedu, R.N.; Li, T.; Zhang, F.; Zhou, H.; Cheng, X.; Jing, X. A systematic investigation of key factors of nucleic acid precipitation toward optimized DNA/RNA isolation. *Biotechniques,* **2020**, *68*(4), 191-199. [http://dx.doi.org/10.2144/btn-2019-0109] [PMID: 32066262]

[4] Ali, N.; Rampazzo, R.C.P.; Costa, A.D.T.; Krieger, M.A. Current nucleic acid extraction methods and their implications to point-of-care diagnostics. *BioMed Res. Int.,* **2017**, *2017*, 1-13. [http://dx.doi.org/10.1155/2017/9306564] [PMID: 28785592]

[5] Tan, SC; Yiap, BC DNA, RNA, and protein extraction: The past and the present. *J. Biomed. Biotechnol.,* **2009**, 574398.

[6] Lee, P.Y.; Costumbrado, J.; Hsu, C.Y.; Kim, Y.H. Agarose gel electrophoresis for the separation of DNA fragments. *J. Vis. Exp.,* **2012**, (62), e3923. [PMID: 22546956]

CHAPTER 14

Introduction to Lipids

Sai Joshi[1], **Sinchan Hait**[2,*] and **Pranali Garde**[3]

[1] *CSIR-NCL, Dr. Homi Bhabha Road, Pune - 411 008, India*

[2] *Department of Chemical Engineering, BITS-Pilani K.K. Birla Goa Campus, Goa-403726, India*

[3] *Stability Data Analytics, Apotex Research Pvt. Ltd., Navi Mumbai, Maharashtra, India*

Abstract: Lipids are water-insoluble hydrocarbon derivatives with many physiological functions. They are the structural components of membranes. The long chain fatty acid chains of lipids can be broken down to serve as energy fuels, as well as they play a role in intracellular signaling. The entire composition of lipid molecules in a cell constitutes its lipidome. In this chapter, we will discuss the different roles played by lipids by their chemical properties in detail.

Keywords: Hydrocarbon, Lipids, Lipidome.

INTRODUCTION

Lipids are a group of natural compounds such as fats, oils, chemicals, and certain parts of membranes that are assembled such that they don't interact with water.

The structure and properties of two delegate lipids are: stearic acid (an unsaturated fat) and phosphatidylcholine (a phospholipid) which are made up of chemical groups that form polar "heads" and nonpolar "tails" (Fig. **14.1**). The polar heads are hydrophilic, or soluble in water, though the nonpolar tails are hydrophobic, or insoluble in water. Lipid particles of this composition immediately form aggregate structures like micelles and lipid bilayers, with their hydrophilic ends arranged toward the watery medium and their hydrophobic ends shielded from the water.

* **Corresponding author Sinchan Hait:** Department of Chemical Engineering, BITS-Pilani K.K. Birla Goa Campus, Goa-403726, India; E-mail: sinchan.hait1@gmail.com

Pamela Jha (Ed.)

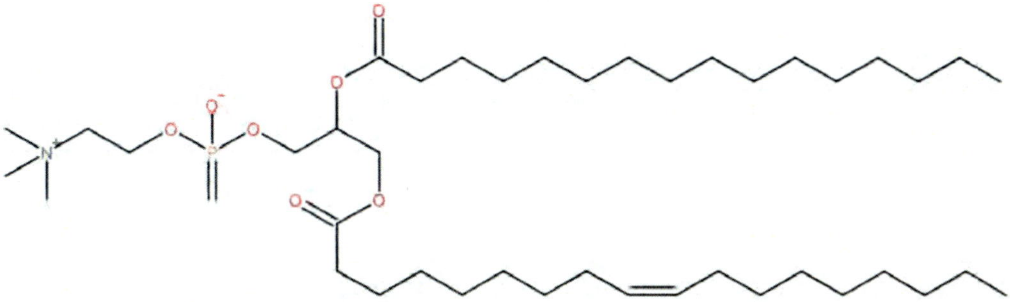

Fig. (14.1). Phospholipid (phosphatidylcholine).

Fats

Fats are any of a group of natural esters of glycerol and different fatty acids, which are solid at room temperature and are the primary constituents of animal and vegetable fat.

Glycerol

Glycerol is a basic polyol compound. Glycerol is found in lipids known as glycerides. Glycerol is an odourless, colourless, viscous liquid that is non-toxic and sweet-tasting.

Fatty acids

Fatty acids consist of a straight chain of a significant number of carbon molecules, with hydrogen atoms along the length of the chain and toward one side of the chain and carboxyl groups (—COOH) at the opposite end. It is the carboxyl group that makes it an acid (carboxylic acid) [1, 2].

Structure of Fatty Acids

Natural fatty acids, individuals from the class of compounds known as carboxylic acids, are made out of a hydrocarbon chain with one terminal carboxyl group (COOH) (Fig. **14.2**). The fragment of a carboxylic acid excluding the hydroxyl (-OH) group is called an acyl group. Under physiological conditions in water, this acidic group typically has lost a hydrogen ion (H^+) to form a negatively charged carboxylate group (-COO). Although the molecule overall is water-insoluble by its hydrophobic hydrocarbon chain, the negatively charged carboxylate is hydrophilic [1, 2]. This normal form for natural lipids, one that contains all-around isolated hydrophobic and hydrophilic parts, is called amphipathic.

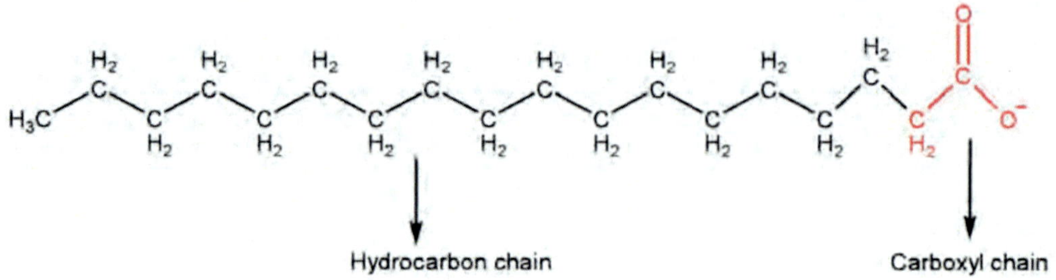

Fig. (14.2). Structural formula for stearic acid.

Classification on the Basis of Saturation

Saturated Fatty Acids

The most straightforward fatty acids are unbranched, direct or linear chains of CH_2 groups connected via a carbon-carbon single bond with one terminal carboxylic acid group [1, 2].

Unsaturated Fatty Acids

Unsaturated fats have at least one or more carbon-carbon double bonds. The term unsaturated shows that less than the most extreme conceivable number of hydrogen atoms are attached to every carbon in the atom. The quantity of double bonds is demonstrated by the conventional name - monounsaturated for atoms with one double bond or polyunsaturated for particles with at least two double bonds. Oleic acid (Fig. **14.3**) is an illustration of a monounsaturated fatty acid.

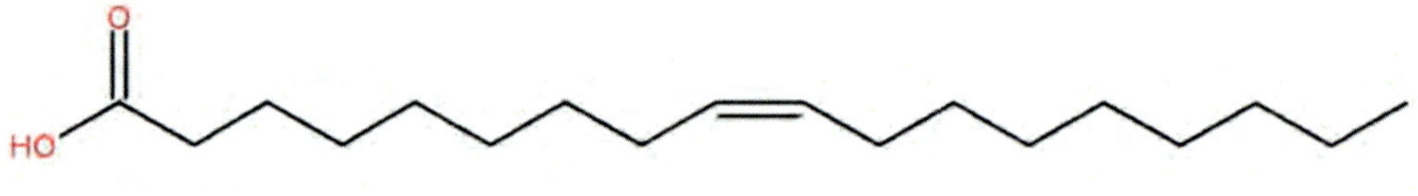

Fig. (14.3). Structural formula of oleic acid.

Fatty Acid Derivatives

Triglycerides

Triglycerides (synthetic name triacylglycerol), the principal method for storing fatty acids in natural frameworks, are a class of mixtures that comprise glycerol (three-carbon trihydroxy alcohol) with a fatty acid connected to every one of the three -OH groups by an **ester bond**. The structure of a typical triglyceride (tristearic acid) is shown in Fig. (**14.4**). Since this molecule contains just one kind of fatty acid, it is alluded to as a simple triglyceride [1, 2].

Waxes

A second group of neutral lipids that are of physiological significance, however, they are a minor part of organic frameworks, are waxes. Waxes consist of a long-chain fatty acid connected through ester oxygen to a long-chain alcohol. These molecules are totally water-insoluble and generally solid at natural temperatures. Their firmly hydrophobic nature permits them to work as water anti-agents on the leaves of certain plants, on feathers, and on the fingernail skin of specific insects.

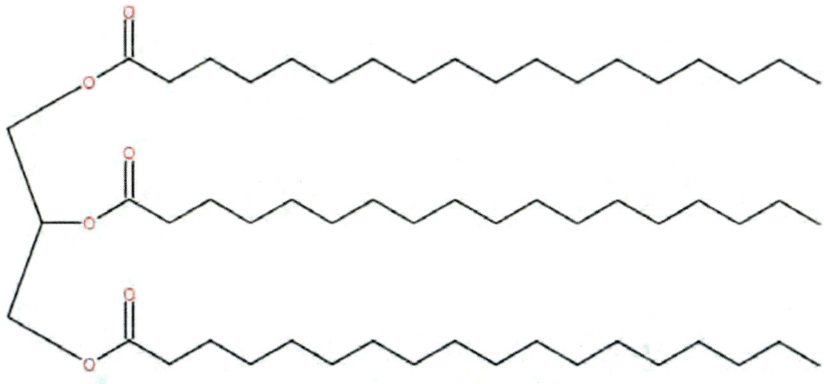

Fig. (14.4). Structural formula of tristearin (tristearic acid).

Glycerophospholipids

Lipids of this class are the most abundant in natural films. In glycerophospholipids (Fig. **14.5**), fatty acids are connected through ester oxygen to carbons 1 and 2 of glycerol, the foundation of the molecule. Phosphate is ester-connected to carbon 3, while any of a few potential substituents is likewise connected to the phosphate moiety. Glycerophospholipids are amphipathic glycerol and phosphate forms the polar end of the particle, while hydrocarbon chains structure the nonpolar end. Although fatty acids can be any of those common in organic frameworks, typically, those attached to carbon 1 are saturated, and those attached to carbon 2 are unsaturated.

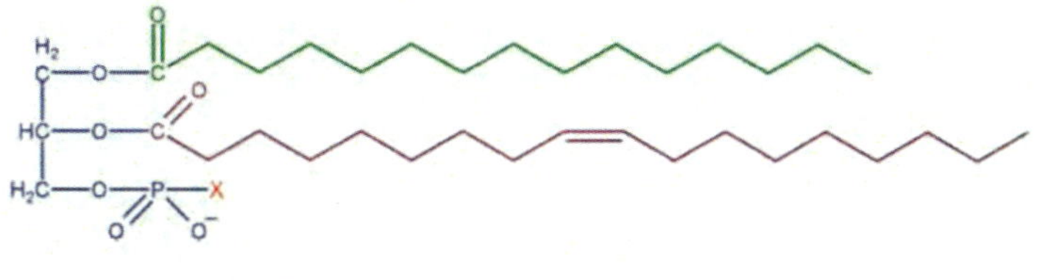

Fig. (14.5). Structure of glycerophospholipids.

Sphingolipids

A subsequent significant class of lipids, for the most part, connected with the films encompassing cells, is sphingolipids. Sphingolipids (Fig. **14.6**) depend on 18-carbon amine alcohol, sphingosine, and to a much lesser degree, on a 20-carbon analogue, phytosphingosine. Everything except one nonexclusive individual from this class has a basic or complex sugar connected to the alcohol on carbon 1.

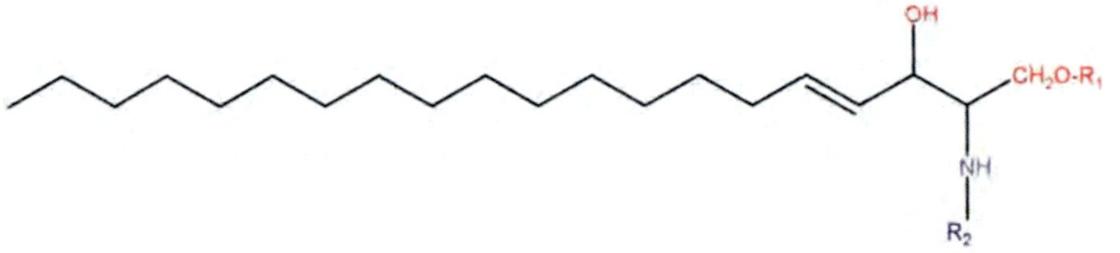

Fig. (14.6). General structural formula for sphingolipids.

Cholesterol and its Derivatives

The sterols are significant parts of natural films in eukaryotes (living beings whose cells have a nucleus) but are uncommon in prokaryotes (cells without a nucleus, like bacteria). Cholesterol (Fig. **14.7**) is the primary sterol of animals, while the significant sterol in fungi is ergosterol and that in plants is sitosterol. The characteristic feature of every one of these three significant molecules is four rigidly fused carbon rings forming the steroid nucleus and a hydroxyl (-OH) group connected to the main ring. One particle is recognized from another by the places of the carbon-carbon double bonds and by the structure of the hydrocarbon side chain on the fourth ring.

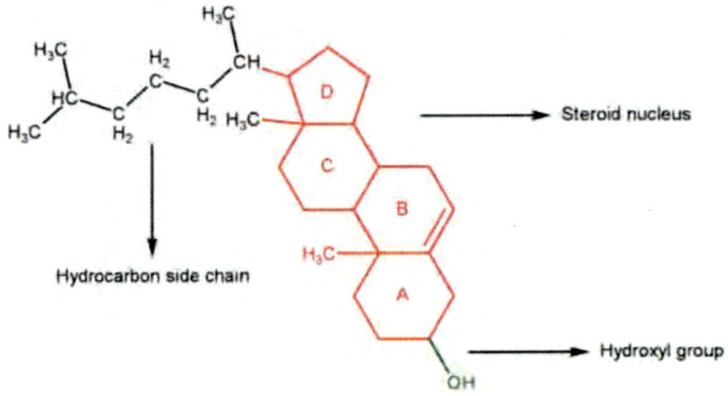

Fig. (14.7). Structural formula for cholesterol.

Lipoproteins

Lipoproteins are lipid-protein complexes that permit all lipids derived from food or synthesized in explicit organs to be transported all through the body by the circulatory system. The essential structure of these totals is that of an oil bead composed of triglycerides and cholesterol esters encompassed by a layer of proteins and amphipathic lipids the same as that of a micelle, a round structure described in the segment Fatty acids.

Classification and Formation

There are four significant classes of circulating lipoproteins, each with its characteristic protein and lipid composition. They are chylomicrons, very low-density lipoproteins (VLDL), low-density lipoproteins (LDL), and high-density lipoproteins (IIDL) [1].

Chylomicrons

Chylomicrons are the biggest lipoproteins, with diameters of 75–600 nm. They have the most minimal protein-to-lipid ratio (being around 90% lipid) and, accordingly the lowest density. Chylomicrons are synthesized by the absorptive cells of the intestinal lining and are discharged by these cells into the lymphatic framework, which joins the blood dissemination at the subclavian vein.

Very Low-Density Lipoproteins (VLDL)

VLDL is a lipoprotein class formed by the liver that is undifferentiated from the chylomicrons discharged by the intestine. Its purpose is to deliver triglycerides, cholesterol esters, and cholesterol to peripheral tissues. VLDL is, to a great extent, exhausted of its triglyceride content in these tissues and leads to an intermediate density lipoprotein (IDL) remnant, which is returned to the liver in the circulation system.

Low-Density Lipoprotein (LDL)

Low-density lipoproteins (LDL) are derived from VLDL and IDL in the plasma and contain a lot of cholesterol and cholesteryl esters. Their foremost job is to deliver these two types of cholesterol to peripheral tissues. Very nearly two-thirds of the cholesterol and its esters found in plasma (blood liberated from red and white cells) are related to LDL.

High-Density Lipoprotein (HDL)

Lipoproteins of this class are the smallest, with a distance across 10.8 nm and the most elevated protein-to-lipid ratio. The subsequent high density gives this class its name. HDL plays an essential part in the removal of excess cholesterol from cells and returning it to the liver, where it is metabolized to bile acids and salts that are ultimately eliminated through the digestive system.

Eicosanoids

Three kinds of locally acting signalling molecules are derived biosynthetically from C20 polyunsaturated fatty acids, basically arachidonic acid. Twenty-carbon fatty acids are completely referred to altogether as eicosanoic acids. The three chemically comparable classes are prostaglandins, thromboxanes, and leukotrienes. The eicosanoids associate with explicit cell surface receptors to create a wide range of outcomes on various tissues, but generally, by and large, they cause provocative reactions and changes in blood pressure, and they additionally influence the coagulation of blood.

Advanced Techniques used for Lipid Quantification

Quantification of lipids can be carried out by two methods:

Conventional Method

Conventional methods for lipid quantification include gravimetric and chromatographic approaches.

Non-conventional Method

Non-conventional methods are based on infrared, Raman, nuclear magnetic resonance, and fluorescence spectroscopic analysis (Fig. **14.8**).

Criteria for Selection of Advanced Method for Lipid Quantification

Gravimetric Analysis

Gravimetric methods for total lipid quantification encompass a series of events including harvesting of microbial biomass, drying, disruption of cellular integrity, and extraction of lipids. With this method, it is only possible to estimate the total amount of lipids, but not their composition.

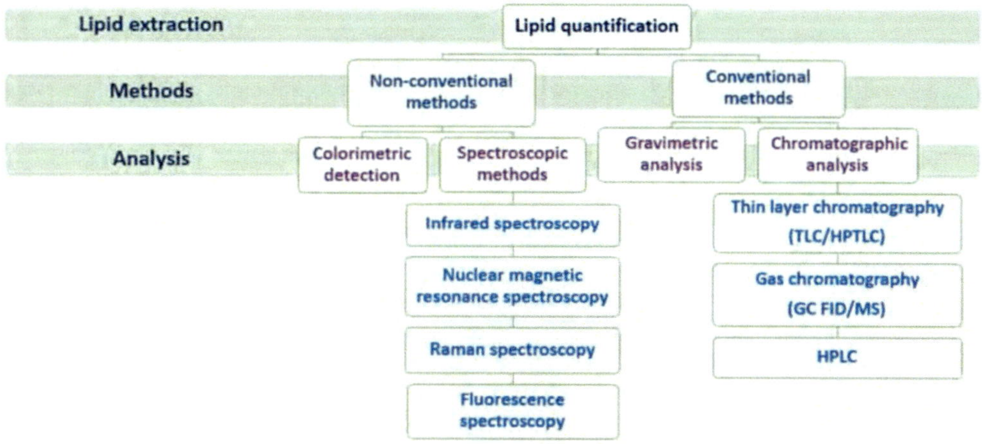

Fig. (14.8). Advanced method for quantification of lipids.

Chromatographic Analysis

Thin-layer Chromatography

Thin-layer chromatography (TLC) is a tried-and-true method for the separation of components in a mixture based on the polarity of the individual components. When a standard is included, this method can also be used for the identification of each component of the mixture.

Gas Chromatography

Gas chromatography (GC) is a highly sensitive method used to identify and quantify the fatty acid content of lipids from tissues, cells, and plasma/serum, yielding results with high accuracy and high reproducibility.

HPLC

The HPLC system that is described is used to separate cholesteryl esters, triglycerides, free fatty acids, and free cholesterol.

Spectroscopic Analysis

Infrared Analysis

FTIR spectroscopy is commonly used in microbial research for the identification and characterization of microorganisms. FTIR spectroscopy is applied to intact biomass and biomass residual after lipid extraction and can serve as a tool for evaluating lipid extraction efficiency.

NMR Analysis

Nuclear magnetic resonance (NMR) spectroscopy is an extremely important technique for the structural analysis and quantification of fatty acids and their derivatives.

Raman Analysis

Raman spectroscopy is a promising tool for determining and quantifying different parameters and properties of lipids.

Fluorescence Analysis

fluorescent staining for lipids offers a rapid and inexpensive analysis tool to measure neutral lipid content.

Colourimetric Analysis

A rapid lipid quantification technique is required to screen the potential lipid-producing strains from the environmental samples. In this study, we used a colourimetric method (SPV) to quantify the total lipids in an aqueous medium.

CONCLUSION

Lipids are water-insoluble compounds present in cells. They can be extracted using non-polar solvents due to their hydrophobic properties. Lipids contain an even number of carbon atoms (12-24) and are classified as saturated or unsaturated. The further classification of fatty acids into triglycerides, waxes, glycerophospholipids, sphingolipids, cholesterol and lipoproteins allows for different physiological functions of lipids. For biochemical analysis of lipids, several techniques are utilized, broadly classified as conventional methods (*e.g.* Gravimetric analysis and chromatographic approaches) and non-conventional methods (*e.g.* Colorimetric and spectroscopy-based methods), which will be discussed in detail in further chapters.

VIVA VOCE

1. Where is HDL synthesized in the human body?

Ans. In the human body, HDL is synthesized in the liver.

2. What are the different techniques that can be used for lipid quantification?

Ans. Lipid quantification can be achieved by gravimetric, chromatophoreic, spectroscopic and colourimetrically.

3. What are the most abundant lipids?

Ans. Glycerophospholipids are the most abundant among other lipids.

4. What is the major difference between saturated and unsaturated fatty acids?

Ans. The major difference between saturated fatty acids and unsaturated fatty acids is that saturated fatty acids have single bonds whereas unsaturated fatty acids have double bonds.

REFERENCES

[1] McNaught, AD I U P A C.*Compendium of chemical terminology,* 2nd; The Gold Book, **1997**.

[2] Subramaniam, S.; Fahy, E.; Gupta, S.; Sud, M.; Byrnes, R.W.; Cotter, D.; Dinasarapu, A.R.; Maurya, M.R. Bioinformatics and systems biology of thc lipidomc. *Chem. Rev.,* **2011**, *111*(10), 6452 6490. [http://dx.doi.org/10.1021/cr200295k] [PMID: 21939287]

CHAPTER 15

Qualitative Tests for Lipid Detection

Sinchan Hait[1,*], **Sai Joshi**[2] and **Pranali Bhosale**[3]

[1] *Department of Chemical Engineering, BITS-Pilani K.K. Birla Goa Campus, Goa-403726, India*

[2] *CSIR-NCL, Dr. Homi Bhabha Road, Pune - 411 008, India*

[3] *Environmental Engineering, Institute of Environmental Engineering, National Sun Yat-Sen University, kaohsiung, Taiwan*

Abstract: Qualitative tests are the primary assays for detecting the presence of any lipids in a given sample. Qualitative detection of lipids is based on the chemical interactions between components of lipids and certain chemical reagents added to them. These chemical reagents allow for the properties of lipids like hydrophobicity, translucency, or colourimetric analysis due to colour changes. These tests have been explained in detail in this chapter.

Keywords: Colourimetric, Hydrophobicity, Lipids, Translucency.

INTRODUCTION

Lipid detection through qualitative tests can confirm the presence of lipids in a mixture. Lipids have unique chemical properties. The solubility of lipid compounds in the mixture employs its detection through qualitative methods. The variation of translucency in mixtures allows for its detection. Under various chemical agents, a colourimetric determination can be correlated with the qualitative determination of lipids.

SOLUBILITY TEST

Aim

Qualitative test for analysis of lipids by solubility test.

* **Corresponding author Sinchan Hait:** Department of Chemical Engineering, BITS-Pilani K.K. Birla Goa Campus, Goa-403726, India; E-mail: sinchan.hait1@gmail.com

Pamela Jha (Ed.)

Principle

This test is used to know the solubility of lipids. Lipids are nonpolar in nature hence they are insoluble in the polar solvent while soluble in nonpolar solvents like chloroform, benzene and boiling alcohol. The general lipid structure of polar and nonpolar lipid molecules is shown below (Fig. **15.1 A-B**) [1 - 3].

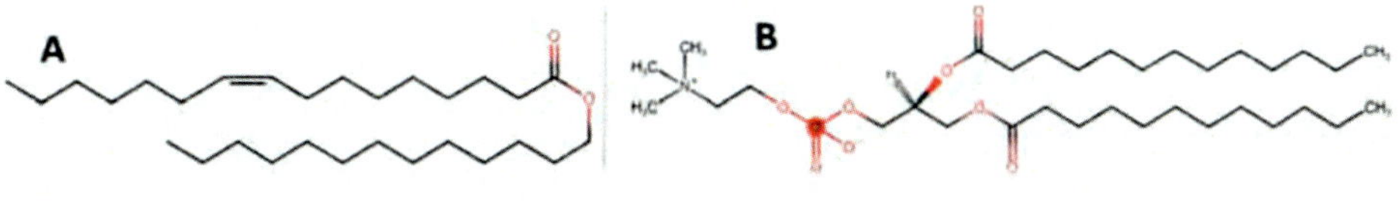

Fig. (15.1). Lipid molecules: **A.** Nonpolar unsaturated **B.** Polar saturated.

Requirements

❖ ***Glassware***:

• Test tubes

❖ ***Chemicals***:

• Olive oil

• Chloroform

• Distilled water

Procedure

1. In two clean, dry test tubes, add 1 ml of olive oil in both test tubes.

2. In the first test tube, mark it as A and add 1 ml of chloroform.

3. In the second tube, mark the test tube as B and add 1 ml of distilled water.

4. Shake both tubes vigorously for 2 minutes.

5. Allow the tubes to stand and note the formation of a homogenous solution with chloroform, indicating that the lipid is dissolved, and the formation of two layers with water indicates that the lipid is insoluble in water.

Observation

1. In test tube A, oil gets dissolved in chloroform.

2. In test tube B, oil drops float on the distilled water surface.

Inference

1. ***Positive result*** Lipids are soluble in a nonpolar solvent, *i.e.*, chloroform and partially soluble in ethanol which can solubilize upon heating.

2. ***Negative result*** Lipids are insoluble in a polar solvent, *i.e.*, water.

The sample contains fat, as it is not soluble in water (test tube B) but soluble only in organic solvents (test tube A).

TRANSLUCENT SPOT TEST

Aim

Qualitative test for analysis of lipids by Translucent Spot test.

Principle

A translucent spot test is also a preliminary test for lipids, which is characterized by a translucent and greasy spot. The lipid will not wet the filter paper, unlike water. The lipids will form a greasy or translucent spot due to their greasy texture, and penetrate the filter paper. Unlike lipids, the spot of water will disappear from the paper [1 - 3].

Requirements

❖ ***Miscellaneous***

• Filter paper

❖ ***Reagents***

• Water

• Olive oil

Procedure

1. Take filter paper.

2. Add 1 ml of water at one end and 1 ml of olive oil at the other end.

3. Observe the appearance of a translucent spot on the filter paper.

Observation

A translucent spot appears at the side where Olive oil is added, while on the side where water is added, no translucent spot is observed.

Inference

1. ***Positive result*** Translucent spots will appear on the filter paper.

2. ***Negative result*** Translucent spot will not appear on the filter paper.

A translucent spot was observed due to the presence of unsaturated fatty acids present in the given sample.

EMULSIFICATION TEST

Aim

Qualitative test for analysis of lipids by Emulsification test.

Principle

An emulsification test is used to detect the presence of lipids. It is the process that stabilizes the water and oil emulsion by using emulsifying agents. The lipid or oil in water will appear as the supernatant. The high surface tension of water develops a separate layer by adding emulsifying agents like bile salts, soap *etc.* Emulsifying agents emulsify the lipid, after which the lipids appear as tiny droplets suspended in the solution [1 - 3].

Keywords: Emulsification, Emulsifying agents

Requirements

❖ ***Glassware***

• Test tubes

❖ ***Chemicals***

• Olive oil

• Soap

Procedure

1. Take two test tubes and label them as test tube A and test tube B. Add 1 ml of water in test tube A, and add 1 ml of water and 1 ml of soap in test tube B.

2. Add 1 ml of olive oil to each of the test tubes.

3. Shake the test tube and allow it to stand for about two minutes.

4. Observe the test tube for the appearance of tiny droplets in the suspension of the liquid.

Observation

Oil droplets are observed in test tubes.

Inference

1. ***Positive result*** It gives a permanent or stable emulsion of lipid and water.

2. ***Negative result*** Oil in water emulsion will form at the top due to the high surface tension of water.

Highly saturated fatty acids give permanent stable emulsion of lipids and water due to their surface tension.

SAPONIFICATION TEST

Aim

Qualitative test for analysis of lipids by Saponification test.

Principle

This test is used to form soap by saponification process (Fig. **15.2**), which includes the hydrolysis of fat or oil with alkaline to yield glycerol and salt of fatty acid (soap) [1 - 3].

Requirements

❖ ***Glassware/Equipment***

- Test tubes
- Water bath

❖ *Chemicals*

- Olive oil
- Alcoholic Potassium hydroxide

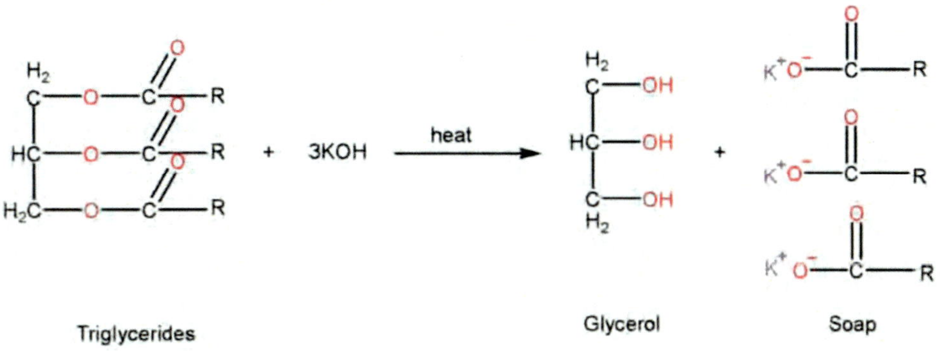

Fig. (15.2). Saponification reaction.

Procedure

1. In clean, dry large test tubes, add 2 ml of olive oil.

2. Add 4 ml of 20% alcoholic potassium hydroxide.

3. Shake well and boil the mixture in a boiling water bath for 5 minutes.

4. Make sure after the five minutes that the soap is formed, all of the fat is converted to soap, and alcohol is evaporated. If not, continue the boiling.

5. The solid material formed is soap.

Observation

Inference

1. ***Positive result*** Froth appears in the test tube.

2. ***Negative result*** Froth does not appear in the test tube.

When the fatty acid possesses a long chain, the salt formed is soap.

SUDAN IV TEST

Aim

Qualitative test for Analysis of Lipids by Sudan IV Test.

Principle

The Sudan IV test is used to detect the presence of lipids in a solution. This test is based on the principle of binding and solubility of lipids in non-polar compounds. As Sudan IV is a non-polar stain, the lipid will bind with it and retain the stain's colour by giving a red-orange colour. Sudan IV does not stain or bind to polar compounds [1 - 3].

Requirements

❖ ***Glassware***

- Test tubes

❖ ***Chemicals***

- Olive oil
- Sudan IV stain
- Water

Procedure

1. Take 1 ml of olive oil in a test tube.
2. Then add 1-2 drops of Sudan IV to the solution.
3. Observe the tubes for the appearance of red-orange colour in the solution.

Observation

Red-orange colour is observed in the test tube containing Olive oil.

Inference

Positive result Gives red-orange colour to the solution.

Negative result The solution of the colour will remain unchanged.

Sudan IV is a highly conjugated aromatic diazo compound, so it is already highly coloured. In the lipid, Sudan IV is hydrophobic, so it will stain any lipid droplets suspended in water.

ACROLEIN TEST

Aim

Qualitative test for analysis of lipids by Acrolein test.

Principle

The acrolein test is used to detect the presence of glycerol and fat. This test is based on the dehydration reaction, in which the water molecules are removed from the glycerol by adding the reagent potassium hydrogen sulphate. The reaction see Fig. (**15.3**) between glycerol and potassium hydrogen sulphate results in acrolein formation, which is characterized physically by the release of a pungent smell [1 - 3].

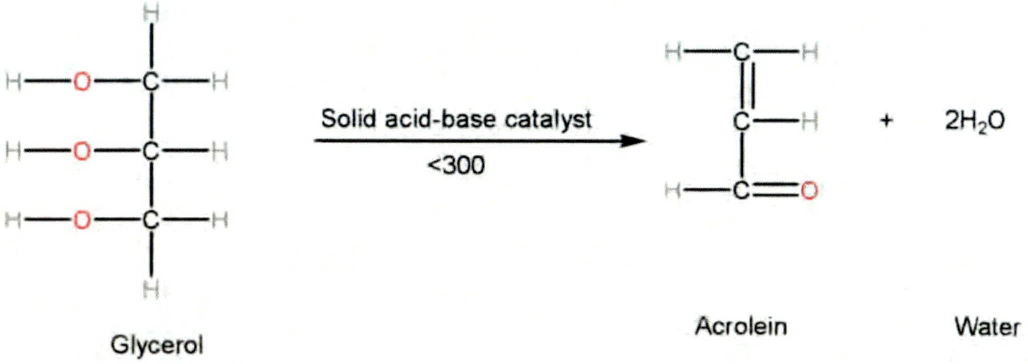

Fig. (15.3). Reaction for formation of Acrolein from Glycerol.

Requirements

❖ ***Glasswares/Equipment***

• Test tubes

• Test tube holder

• Burner

❖ *Chemicals*

- Potassium hydrogen sulphate

Procedure

1. Take 1 ml of olive oil in a test tube.
2. Add crystals of potassium hydrogen sulphate.
3. Heat the solution for a few minutes.
4. Smell the test tube for the pungent smell.

Observation

A pungent smell was observed from the test tube containing olive oil and potassium hydrogen sulphate.

Inference

1. ***Positive result*** If glycerol is present in the sample, it will give a pungent smell.
2. ***Negative result*** If glycerol is absent in a sample, it will not produce a pungent smell.

The reaction between glycerol and potassium hydrogen sulphate gives a pungent smell.

DICHROMATE TEST

Aim

Qualitative test for analysis of lipids by Dichromate test.

Principle

The dichromate test is also used to detect the presence of glycerol. It is based on the principle of an oxidation reaction. In this, glycerol and dichromate ions react to give a brown colour to the solution. Then, the chromic ions oxidize the glycerol and reduce into chromous ions by giving a blue colour to the solution in the presence of nitric acid [1 - 3].

Requirements

❖ ***Glassware/Equipment***

- Test tubes

❖ ***Chemicals***

- Olive oil
- 5% Potassium dichromate solution
- Nitric acid

Procedure

1. Take 2-3 ml of olive oil in a test tube.
2. Then, add a few drops of 5% potassium dichromate solution.
3. After that, add 5 ml of concentrated nitric acid.
4. Observe the test tube for the appearance of a blue colour.

Observation

The blue colour appears in the test tube containing olive oil.

Inference

1. ***Positive result*** If the colour of the solution changes from brown to blue, it indicates glycerol.
2. ***Negative result*** In this, the brown colour will not change into blue.

The reaction between Olive oil, Potassium dichromate solution, and Nitric acid solution gives a blue colour to the solution.

TEST FOR FREE FATTY ACIDS

Aim

Qualitative test for analysis of lipids by testing for free fatty acids.

Principle

This is based on the neutralization reaction, where the alkali neutralises by adding free fatty acids into the lipids [1 - 3].

Requirements

❖ ***Glassware/Equipment***

• Test tube

❖ ***Chemicals***

• Phenolphthalein solution

• Olive oil

Procedure

1. Add phenolphthalein solution to a test tube.

2. Then, add dilute alkali to the above solution (gives a pink colour).

3. At last, add 1 ml olive oil.

4. Observe the tube for the disappearance of the pink colour after the addition of lipids.

Observation

The pink colour disappeared from the test tube containing olive oil.

Inference

The disappearance of the pink colour from the test tube shows the presence of lipids in it. This is because the neutralization reaction takes place.

Troubleshooting

1. Positive result: If the pink colour disappears by adding the lipid sample, it indicates free fatty acids in the sample.

2. Negative test: The pink colour will not disappear.

UNSATURATION TEST

Aim

Qualitative test for analysis of lipids by Unsaturation test.

Principle

The unsaturation test is used to detect the unsaturated fatty acids or double bonds in a lipid sample. All the neutral fat contains glycerides of fatty acids. Double bonds are found in the structure of unsaturated fatty acids, which become saturated by taking up either bromine or iodine (Fig. **15.4**). If the lipid contains more unsaturated fatty acids or more double bonds, it will take more iodine [1 - 3].

Requirements

❖ *Glassware/Equipment*

- Test tube
- Beaker
- Dropper

❖ *Chemicals*

- Chloroform
- Huble's iodine reagent

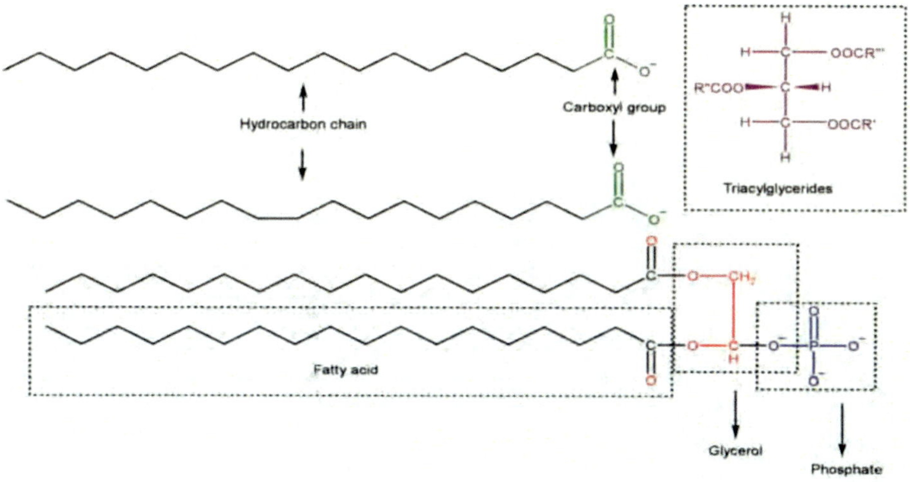

Fig. (15.4). Showing saturated and unsaturated lipids.

Procedure

1. Take 5 ml of chloroform and 5 ml of Huble's iodine reagent in a beaker, giving pink colour to the solution.

2. Add 1 ml of olive oil drop by drop and shake vigorously, until the pink colour disappears.

3. Count the number of drops added to chloroform and Huble's iodine solution until the pink colour disappears. The number of drops determines the taking up of iodine by the unsaturated fatty acid of lipids.

Observation

The pink colour disappeared from the test tube containing olive oil.

Inference

1. ***Positive result*** Pink colour will disappear with the addition of unsaturated fatty acids.

2. ***Negative result*** Pink colour will not disappear.

These unsaturated fatty acids become saturated by taking up iodine.

BURCHARD TEST

Aim

Qualitative test for analysis of lipids by Burchard test.

Principle

The Burchard test was first given by the scientist Liebmann to detect the presence of cholesterol. Cholesterol reacts with the strong concentrated acid, *i.e.*, sulphuric acid and acetic anhydride (Fig. **15.5**). Sulphuric acid and acetic anhydride act as a dehydrating and oxidizing agent [1 - 3].

Requirements

❖ ***Glassware/Equipment***

- Test tube
- Dropper

❖ *Chemicals*

- Crystals of Cholesterol
- Chloroform
- Acetic anhydride
- Conc. sulphuric acid

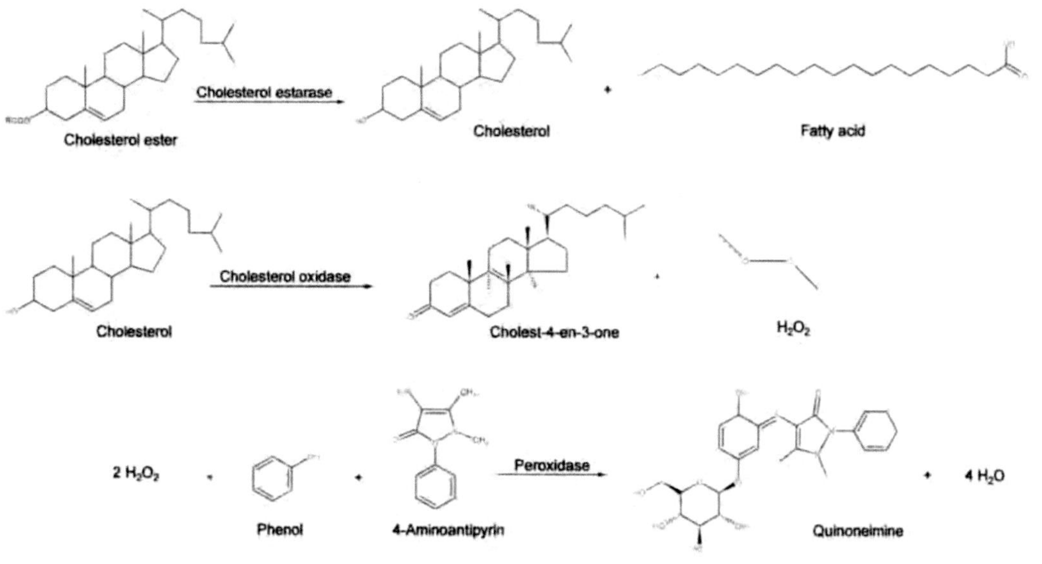

Fig. (15.5). Reaction for Burchard test.

Procedure

1. Take crystals of cholesterol in a test tube.
2. Then, add 2 ml of chloroform to dissolve the cholesterol.
3. Add 10 drops of acetic anhydride in a solution and 2-3 drops of concentrated sulphuric acid.
4. Observe the test tube for the appearance of a bluish-green colour.

Observation

Bluish-green colour is observed in the test tube containing Cholesterol crystals.

Inference

1. ***Positive result*** It indicates cholesterol in a sample by giving bluish-green colour to the solution.

2. ***Negative result*** The colour of the solution will not change.

Cholesterol reacts with strong acids, which results in a change in the colour of the solution to a bluish-green colour.

Precautions

1. Minimize the air current in the working area so it is possible to avoid spores of contaminating microorganisms moving in along with the air currents over the sterile areas. At least the fan should not be used in the laminar airflow bench room (inoculation room). Preferably, all the places should be air-conditioned.

2. Store properly the prepared media, nutrients and tools in cabinets.

3. Use a separate area for cleaning and washing and the preparation of the medium.

4. UV tubes are also fixed in the Laminar air flow bench placed in the inoculation chamber. All these UV tube lights should be used frequently before inoculations. UV lights in corridors may be left open at night.

DID YOU KNOW?

1. Fat enables the absorption of some nutrients including vitamins A, D, E, and K.

2. Fats have more than two times the quantity of energy as protein and carbohydrates. (1 gram of protein = 4 Kilocalories, 1 gram of carbohydrate = 4 Kilocalories, 1 gram of fat = 9 Kilocalories).

3. Fat consumption must no longer exceed 30% of your general energy. To calculate grams of fats for an average 2000 calories diet:

2000 energy X 30% = 600 fats energy

600 fats energy ÷ 9 energy in step with grams of fats = 66 grams of fats/day.

4. Abnormal metabolism of lipids and lipoproteins has been strongly related to disorders, like type 2 diabetes mellitus, cardiovascular diseases, and cancers.

5. Waxes encompass a long-chain fatty acid connected thru an ester of oxygen to

a long-chain alcohol. These are observed in ear canals, leaves, and beeswax which makes honeycombs. Waxes shape a barrier to exclude water in each animal and plant.

6. Steroids are needed to make each of the female (estrogen) and male (testosterone) hormones.

CONCLUSION

In this chapter, several qualitative tests are discussed. Based on the chemical properties of the lipids, they react with certain chemical components and produce some colour compounds or display physical appearance-based characteristics, which then can be interpreted as the presence of lipids in the sample. However, these techniques can be used for primary analysis purposes only. Further, quantifying carbohydrates needs to be done, which is discussed in the next chapter.

VIVA VOCE

1. Describe the structural difference between saturated and unsaturated fatty acids. What are polyunsaturated fatty acids?

Ans. In saturated fatty acids, every carbon has 4 hydrogens bonded to it, which offers it an immediate shape and subsequently lets them stack on the pinnacle of every different (Fig. **15.6**). As they stack, they're solid at room temperature. Unsaturated fat has, as a minimum, one double bond with any other carbon. This double bond offers the molecule an unethical shape, subsequently cannot stack on a pinnacle of every different and are liquid at room temperature. Polyunsaturated fatty acids (PUFAs) are fatty acids that include a couple of double bonds in their structure.

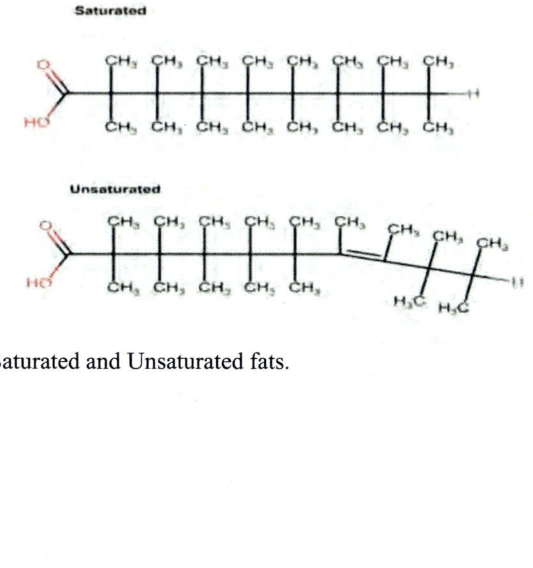

Fig. (15.6). Structure of Saturated and Unsaturated fats.

2. Which chemical reaction differentiates between unsaturated and saturated fatty acids?

Ans. Unsaturated fatty acids are much less solid than saturated fatty acids, which makes them at greater risk of rancidity. Rancidity is the oxidation of fat that is due to hydration, oxidation, steel atoms or microbes and commonly results in a foul smell and/or taste.

3. Describe at least 3 important functions for cholesterol in the body.

Ans. The main functions of cholesterol are as folows:

- Maintain the fluidity and integrity of cell membrane.
- Serve as a precursor for synthesis of steroid hormones, bile acids and Vitamin D.
- It assists in the manufacturing of bile within the liver.

4. What are trans fatty acids? Which foods are the sources of fatty acids?

Ans. The synthetic shape of trans fat, called in part hydrogenated oil, can be located in a whole lot of meals products, together with baked goods, which includes cakes, cookies and pies; Refrigerated dough, which includes biscuits and rolls; Fried foods, together with french fries, doughnuts and fried chicken, non-dairy espresso creamer *etc.*

5. Many processed foods are labelled "Cholesterol free". From what food source would these foods be made? Animal or Vegetable?

Ans. Cholesterol is simplest determined in meals that come from animals, there may be no LDL cholesterol in meals that come from plants. So, there may be no LDL cholesterol in fruit, vegetables, grains, seeds, nuts, beans, peas and lentils

REFERENCES

[1] Tiwari, A. *Practical Biochemistry: A Student Companion*; LAP LAMBERT Academic Publishing, **2015**.

[2] Mu, P.; Plummer, D.T. *Introduction to practical biochemistry*; Tata McGraw-Hill Education: Noida, **2001**.

[3] Sawhney, S.K. *Introductory practical biochemistry*; Singh, R., Ed.; Alpha Science Int'l Ltd., **2000**.

CHAPTER 16

Quantitative Tests for Lipid Detection

Sai Joshi[1], **Sinchan Hait**[2,*] and **Pranali Garde**[3]

[1] *CSIR-NCL, Dr. Homi Bhabha Road, Pune - 411 008, India*

[2] *Department of Chemical Engineering, BITS-Pilani K.K. Birla Goa Campus, Goa-403726, India*

[3] *Stability Data Analytics, Apotex Research Pvt. Ltd., Navi Mumbai, Maharashtra, India*

Abstract: Quantification of lipids can be broadly classified into the following methods: gravimetric analysis, which employs the principle of phase separation, the acid value is based on the property of rancidity of fats, saponification value which relies on alkali required to saponify (hydrolyze) lipids containing mixture, iodine value depends on the interaction of fats with halogen iodine, and blood cholesterol estimation determines the cholesterol level by cholesterol esterase enzyme-based assay. These assays have been explained in detail in this chapter. Advanced techniques used for lipid quantification utilize the principles of chromatography (Thin-layer chromatography, gas chromatography, and HPLC) and spectroscopy (infrared spectroscopy, Raman spectroscopy, nuclear magnetic resistance spectroscopy, and fluorescence spectroscopy), and different colourimetric assays are discussed in detail in this chapter.

Keywords: Cholesterol, Fluorescent spectroscopy, Gravimetric, Gas chromatography, HPLC, Iodine value, Lipid, Raman spectroscopy, Thin-layer chromatography.

INTRODUCTION

Qualitative analysis of lipids allows for the detection of lipids in a mixture, however, quantification of the same is an important step. Over the period, several techniques have been developed; however, there are some standard assays available which are performed routinely. Manipulation of physical and chemical characteristics of lipids to estimate the number of lipids in a mixture allows us to understand the number of lipids in the mixture. The importance of lipid determination to maintain proper physiological functions is an important application of these quantitative techniques. Lipids like cholesterol and fats play an important role in the physiological functioning of the cells, and it ultimately

* **Corresponding author Sinchan Hait:** Department of Chemical Engineering, BITS-Pilani K.K. Birla Goa Campus, Goa-403726, India; E-mail: sinchan.hait1@gmail.com

Pamela Jha (Ed.)

impacts the entire composition of human functioning. Using quantitative techniques, a deeper understanding of lipid composition can be obtained. In this chapter, we have discussed different quantitative measures to estimate the number of lipids in a mixture.

GRAVIMETRIC ANALYSIS

Aim

Estimation of total lipids by the Bligh and Dyer method.

Principle

Methanol, chloroform and water are added to the sample in a two-step extraction and after phase separation, lipids are quantified in the chloroform phase. The mixtures with a known measure of water (80%) are broken down in this binary system and are then isolated by the transition from the monophasic system to the biphasic system induced by the addition of water: the final composition is situated inside the miscibility gap: the two samples with compositions given by the points at the end of the tie-line are coexist [1 - 5].

Requirements

❖ ***Reagents***

- Chloroform
- Methanol

❖ ***Other requirements***

- Homogenizer
- Whatman Filter Paper
- Plastic vials
- Measuring Cylinders
- Aluminium pans

Procedure

1. 5 g samples (preferably dry wet weight) are homogenized for 2 min in a homogenizer with 5 ml each of chloroform and methanol.

2. The mixture is homogenized for another 30 s after the addition of 5 ml of chloroform.

3. 5 ml of water is added to the mixture, and the sample is homogenized again for 30 s.

4. The mixture is then allowed to separate, initially, the lower solvent phase is removed and then passed through a Whatman #1 filter paper, and then the filtrate is saved in a labelled vial.

5. Another 5 ml of chloroform is added into the remaining pellet and aqueous phase and homogenized for another 2 min.

6. The resultant mixture is added to the previous filtrate by passing it through the Whatman #1 filter paper.

7. The filtrate is allowed to separate in the graduated cylinder, and the volume of the lower chloroform layer is recorded.

8. The lipids are then gravimetrically determined by placing 0.5 ml aliquots of the chloroform layer into pre-weighed aluminum pans (three pans per sample), allowing the samples to evaporate in a hood overnight, recording the weights, and then converting to per cent lipids.

Observation

Determination of Total Lipid Content

1. After weighing, a small volume of chloroform is added to each flask to detect the presence of non-lipid material that is insoluble.

2. If non-lipids are present, the chloroform is carefully decanted, and the flask is rinsed 3 times with the chloroform.

3. Then, the dry weight of the residue is determined and subtracted from the initial weight.

4. The lipid content of the sample is calculated as follows:

$$Total\ Lipid\ (g) = \frac{Weight\ of\ lipid\ in\ aliquot\ X\ Volume\ of\ chloroform\ layer}{Volume\ of\ aliquot}$$

Results

In general, lipid extraction from plants can be performed through two different routes, the dry and wet routes. The extracted lipids are often contaminated with other cellular components, which affects their determination and creates a discrepancy between gravimetric and fatty acid methyl ester (FAME) analysis.

Precaution

1. Precipitation ought to be practically complete and free from contamination.

2. Equilibrium ought to be adjusted carefully.

3. Grinding the sample is compulsory.

Application

1. It determines the purity and thermal stability of both primary and secondary standards.

2. It also helps in determining the composition of the complex mixture.

3. It helps in studying the sublimation behaviour of various substances.

Advantages

1. It provides for exceedingly precise analysis.

2. It can be used to determine the atomic masses of various elements.

3. Gravimetry provides negligible instrumental error and does not require a series of standards for calculations of unknown samples.

4. This method often does not require expensive equipment.

Disadvantages

1. Gravimetric analysis usually gives an analysis of a single element, or a limited group of elements, at a time.

2. But there are potential problems with gravimetric analysis that must be avoided to get good results.

3. Proper lab technique is critical.

4. This method is relatively time-consuming.

ACID VALUE

Aim

To determine the acid value in oil and fats.

Principle

Fats are readily broken down by lipase into free fatty acids and glycerol during storage, especially when the temperature and moisture contents are high (Fig. **16.1**). Fat hydrolysis gets sped up because of microbial contamination [1 - 3].

$$RCOOH + KOH \longrightarrow RCOO^-K^+ + H_2O$$

RCOOH	KOH	RCOO⁻K⁺	H₂O
Carboxylic acid	Potassium hydroxide	Soap	Water

Fig. (16.1). Formation of soap.

Requirements

❖ ***Sample***

Fat or oil

❖ ***Reagent***

- 0.1 M KOH
- Ethanol
- Diethyl ether
- 1% Phenolphthalein solution

❖ ***Apparatus***

- Pipettes
- Conical flask
- Burette
- Burette stands

Procedure

1. Blend the oil or melted fat completely before weighing. The mass of the test will be taken based on the colour and expected acid value, as shown in Table **16.1**:

Table 16.1. Expected acid value based on the colour of the test.

Expected Acid Value	Mass of Test Portion (gm)	Accuracy of Weighing of Test Portion (gm)
< 1	20	0.005
1 to 4	10	0.020
4 to 15	2.5	0.010
15 to 75	0.5	0.001
> 75	0.1	0.0002

2. Weigh precisely in an appropriate amount of the cooled oil test, as mentioned in the above table, in a 250 mL conical flask.

3. Add 50 mL of freshly neutralised hot ethyl alcohol and around 1 ml of phenolphthalein indicator solution.

4. Heat the solution mixture for around 15 minutes in a water bath (75-80°C).

5. Titrate while hot against standard alkali solution shaking vigorously during the titration.

6. The endpoint utilizing the phenolphthalein indicator will be from colourless to light pink (Persisting for 15 sec.)

7. The endpoint utilizing Alkali blue 6B pointer will be the vanishing of the blue colour which developed during the addition of the indicator.

8. The weight of the oil/fat taken for the estimation and the strength of the alkali utilized for titration will be such that the volume of alkali needed for the titration doesn't surpass 10mL.

Calculation

$$Acid\ value = \frac{56.1\ x\ V\ x\ N}{W}$$

Where,

V = Volume in mL of standard potassium hydroxide or sodium hydroxide used.

N = Normality of the potassium hydroxide solution or Sodium hydroxide solution.

W = Weight in gm of the sample.

Result

The acid value of a given fat or oil sample is found to be__________g/ml.

Precaution

1. Observing the burette readings after acquiring dark pink colour OR Orangish red as an endpoint ought to be avoided as it will lead to false results.

2. In the case of rice bran oil or RBO-based samples, add 1mL of Alkali blue indicator instead of phenolphthalein indicator for avoiding false results.

Application

As rancidity is normally accompanied by free fatty acid development, the determination of acid value is regularly utilized as an overall indication of the condition and edibility of oils.

Advantages

1. There is no presence of a toxic solvent.

2. Extraction and titration of acids are done at room temperature.

3. In this method, there is no need for preliminary neutralization of acid admixtures in a solvent.

Disadvantages

Disadvantages of this method and comparable strategies are the utilization of organic solvents (volume and toxicity), the requirement for heating the reaction media, inadequate solubility of the oil/fat, the need to pre-neutralize the solvents, utilization of a lot of tests, and the chance of mistake to recognize the colour change of the indicator analyzing coloured samples.

SAPONIFICATION VALUE

Aim

To estimate the Saponification value of oils.

Principle

Saponification esteem is characterized as the number of milligrams of KOH needed to completely hydrolyse (saponify) one gram of the oil/fat. A known measure of the oil or fat is refluxed with an excess measure of standard alcoholic potassium hydroxide solution and unused alkali is titrated against a standard acid solution see Fig. (**16.2**) [1 - 4].

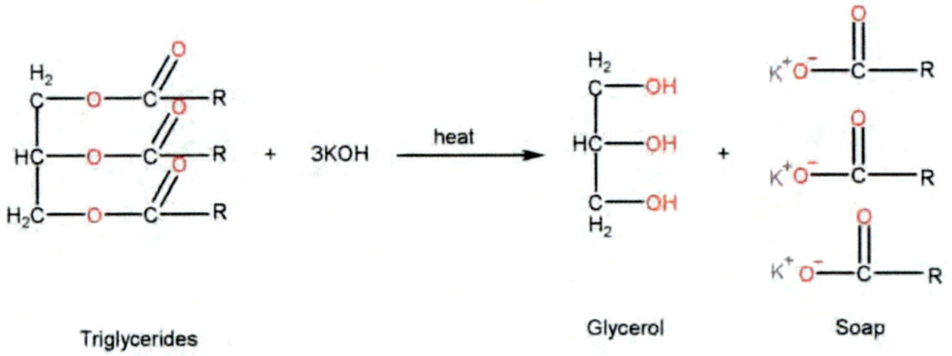

Fig 16.2. Saponification of fats.

Requirements

❖ Sample:

oil or fat

❖ ***Apparatus***

- Conical flask
- Reflux and condenser
- Water bath
- Burette and Pipette

❖ ***Chemicals***

- 0.5 N HCl
- Phenolphthalein
- 0.5 N alcoholic potassium hydroxide solution

Procedure

1. Around 2 g of the given oil/fat is taken in a conical flask, however weighed precisely.

2. The oil or fat is disintegrated in 25 ml of 0.5 N alcoholic potassium hydroxide solution.

3. Then, at that point, the reaction mixture is refluxed utilizing a water condenser on a water bath for 30 minutes.

4. The resulting solution is then cooled and titrated against a 0.5 N HCl solution, and then add 1 ml of phenolphthalein.

5. Proceed with the titration until the colour of the solution in the flasks turns pink to white.

6. Note down the endpoint of both the blank as (a) and the test as (b).

Calculation

$$Saponification\ value = \frac{Volume\ of\ acid\ required\ to\ neutralize\ KOH\ X\ Equivalent\ factor\ x\ 1000}{W}$$

$$= \frac{(a-b)\ X\ 0.02805\ X\ 1000}{W}$$

Result

The saponification value of given fat or oil is found to be___________ml.

Precaution

1. Care ought to be taken while dealing with reagents like potassium hydroxide and hydrochloric acid. Accidental spills of these reagents will cause a serious Itching sensation. Wash the affected area with cold water and inform the lab assistant right away.

2. Caution ought to be taken while attaching the reflux condensers to the conical flask.

3. Ensure that the water bath is set to 100°C and the reflux condensers are set with legitimate settings before the beginning of experiments.

Application

Saponification is utilized by wet chemical fire extinguishers to convert burning fats and oils into non-combustible soap, which helps in extinguishing the fire.

Advantages

1. Saponification is essential to the industrial user to know the measure of free fatty acids in a food material.

2. The amount of free fatty acids can be recognized by determining the amount of alkali that should be added to the fat or oil to make it neutral.

Disadvantages

The end-point in the butyl alcohol system isn't as sharp as the typical saponification end-point in ethyl alcohol. This might be brought about by the two-phase nature of the titration mixture.

Did You Know?

Phenolphthalein was utilized broadly in over-the-counter laxatives, however, in 1999, the U.S. Food and Drug Administration restricted its utilization of such medicines since animal studies showed that they might cause cancer in humans.

IODINE VALUE

Aim

To determine the iodine value of fats and oils and thus estimate the unsaturation of the fats and oils.

Principle

Fatty acid reacts with the halogen (Iodine) at the C=C double bond site. In this reaction, iodine monochloride reacts with the unsaturated bond to deliver a di-halogenated single bond, of which one carbon has bound an atom of iodine (Fig. **16.3**) [1 - 4].

Fig. (16.3). Formation of a di-halogenated single bond compound.

After the reaction is finished, the amount of iodine that has reacted in the reaction is determined by adding potassium iodide solution to the reaction product (Fig. **16.4**).

$$ICl + KI \longrightarrow KCl + I_2$$

Fig. (16.4). Formation of Potassium chloride and Iodine.

This causes the excess unreacted ICl to produce molecular iodine. The produced I_2 is then titrated with a standard 0.1N sodium thiosulfate solution (Fig. **16.5**).

$$I_2 + 2Na_2S_2O_3 \longrightarrow 2NaI + Na_2S_2O_4$$

Fig. (16.5). Formation of sodium iodide and sodium dithionite.

Saturated fatty acid won't give the halogenation reaction. If the iodine number is between 0-70, it will be fat and if the value surpasses 70, it is an oil. Starch is utilized as the indicator for this reaction so the liberated iodine will react with starch to give a purple-coloured complex and along these lines, the endpoint can be observed.

Requirements

❖ *Sample*

Fat sample in chloroform

❖ *Reagents*

- Iodine Monochloride Reagent
- Potassium Iodide Standardized
- 0.1 N Sodium thiosulphate
- 1% Starch indicator solution
- Chloroform
- Distilled water

❖ ***Glasswares***

- Reagent bottle
- Iodination flask
- Burette and burette stand with a magnetic stirrer
- Glass pipette
- Measuring cylinder

Procedure

1. Arrange all the prepared reagent solutions and the necessities on the table.

2. Pipette out 10 ml of fat sample disintegrated in chloroform to an iodination flask marked as TEST.

3. Add 20 ml of Iodine Monochloride reagent into the flask. Mix all the requirements in the flask completely.

4. Then, the flask is permitted to stand for 30 minutes of incubation in the dark.

5. Set up a BLANK in another iodination flask by adding 10ml Chloroform to the flask.

6. Add to the BLANK, 20 ml of Iodine Monochloride reagent and mix the solutions in the flasks completely.

7. Incubate the BLANK in the dark for 30 minutes.

8. Meanwhile, take out the TEST from incubation after half an hour and add 10 ml of potassium iodide solution into the flask.

9. Wash the plug and the sides of the flask utilizing 50 ml of distilled water.

10. Titrate the TEST against standardized sodium thiosulphate solution until a pale straw colour is observed.

11. Add about 1ml starch indicator into the solutions in the flask, and purple colour is observed.

12. Proceed with the titration until the colour of the solution in the flasks turns colourless.

13. The vanishing of the blue colour is observed as the end point of the titration.

14. Also, the technique is repeated for the BLANK.

15. Record the endpoint readings of the BLANK.

Calculation

$$Volume\ of\ Sodium\ thiosulphate\ used = (Blank - Test)\ ml$$

$$Iodine\ no. of\ fat = \frac{Equivalent\ Wt. of\ iodine\ x\ Volume\ of\ NaS\ x\ Normality\ of\ Na2S2O3\ x\ 100x\ 10 - 3}{Weight\ of\ fat\ sample\ used\ for\ analysis\ (g)}$$

Where,
The equivalent weight of iodine = 127
Normality of sodium thiosulphate ($Na_2S_2O_3$) = 0.1

Results

The iodine value of fats is found to be___________ml.

Precaution

1. Iodine monochloride is harmful. So, handle the reagent with gloves.

2. For better outcomes, perform the tests without any time gap during the addition of reagents, as the freed liberated is vulnerable to oxidation by light.

Application

The most important application of the iodine value is to determine the measure of unsaturation contained in fatty acids.

Advantages

Since the melting point and oxidative stability are similar to the degree of unsaturation, the Iodine Value (IV) gives an estimation of these quality elements. The more noteworthy the iodine value, the more unsaturation and the higher the susceptibility to oxidation.

Disadvantages

One of the significant limitations is that halogens don't respond stoichiometrically with conjugated double bonds (especially abundant in some drying oils).

Did You Know?

Iodine is required for the production of thyroid hormones in humans as well as in all vertebrates. Its deficiency can lead to goitre (broadened thyroid gland), intellectual disability and cretinism.

BLOOD CHOLESTEROL

Aim

Determination of cholesterol level in human serum - Enzymatic Colourimetric Method.

Principle

The cholesterol esters of the sample are hydrolyzed by cholesterol esterase. 4-Cholesten3-one and H_2O_2 are then produced from the released free cholesterol by cholesterol oxidase. A quantifiable red quinone imine derivative, that has an absorbance at 500 nm, is produced from hydrogen peroxide (H_2O_2) and 4-amin--antipyrine in the presence of phenol and peroxidase (Fig. **16.6**) [1 - 5].

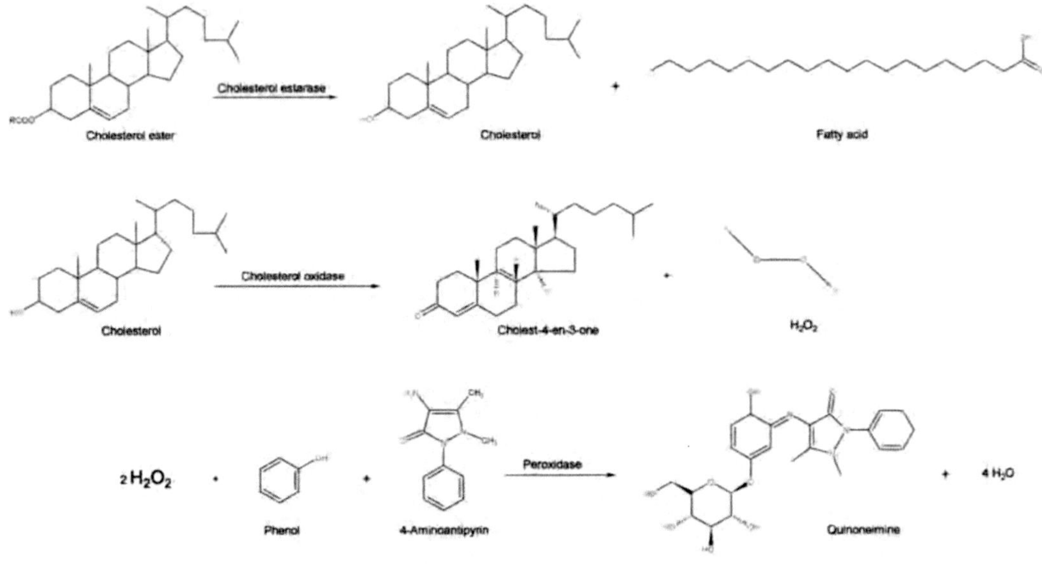

Fig. (16.6). Schematic representation of the reaction.

Requirements

❖ *Sample*

Human serum sample

❖ *Reagent*

- Cholesterol buffer solution
- Enzyme reagent
- Cholesterol standard solution - 300mg/dL
- Distilled water

❖ *Apparatus*

- Water bath
- Test tubes
- UV/Vis Spectrophotometer Cuvettes (10 mm pathlength)

Procedure

1. Prepare the enzyme solution by disintegrating the enzyme reagent to 100 mL in the cholesterol buffer solution.

2. Prepare the solution mixture as displayed in the table.

3. Keep the tubes in a water bath at 37 °C for 5 minutes.

4. Measure blank solutions, standard solutions and tests using UV/Vis Spectrophotometer (Table **16.2**).

Table 16.2. Mixture preparations.

-	Blank	Standard	Sample
Enzyme solution	3	3	3
Distilled water	0.02	-	-
Cholesterol standard	-	0.02	-
Serum sample	-	-	0.02

Normal Range: 130 - 250 mg/dL

Calculation

$$Cholesterol\ level\ (mg/dl) = \frac{Absorbance\ of\ sample}{Absorbance\ of\ standard}\ X\ 300$$

Results

The cholesterol level in the human serum sample is found to be___________mg/dl.

Precaution

1. The material and drying are major factors in the reagent stability and obtaining correct outcomes.

2. The water in the lab to prepare the reagents and use in the estimations should be distilled water.

Advantages

1. This method requires a smaller sample quantity.

2. It is more Rapid; does not require any preliminary extraction step.

Disadvantages

1. This method is not specific to cholesterol.

2. In this method, Ascorbic acid and Bilirubin are interfered with by consuming H202, and thus, Bilirubin interference can produce falsely high or low values.

Did You Know?

1. High cholesterol has no symptoms, so many individuals don't realize that their cholesterol is excessively high.

2. Having high blood cholesterol raises the danger of heart disease, the leading cause of death, and for stroke, the fifth leading reason for death.

ADVANCED TECHNIQUES OF LIPID QUANTIFICATION

Chromatographic Analysis

Thin Layer Chromatography

Aim

Lipid Component Quantitation by Thin Layer Chromatography.

Principle

As TLC is a chromatographic method, it comprises a stationary phase, a thin layer of adsorbent (normally silica) coated on a glass support/plate, and a mobile phase, commonly consisting of at least two natural solvents. Samples are spotted close to the bottom part of the TLC plate (assigned as the "origin"), which is then placed in a developing chamber containing a shallow layer of solvent. The solvent (the mobile phase) moves up the TLC plate because of capillary action. The segments of the sample migrate up the TLC plate as indicated by their solidarity of interaction with the mobile and stationary phases. When the solvent has reached the top point of the plate (alluded to as the "solvent front"), the plate is taken out from the developing chamber and permitted to dry. The spots (*i.e.*, the isolated parts of the example) are then visualized by oxidation and thermal charring. Quantitation is then performed by image analysis of the charred sample in contrast with calibration standards [1 - 5].

Application

1. Separation of different lipid classes from animal and plant tissues.

2. Determine the number of components in a mixture to determine the identity of two substances, or monitor the progress of a reaction.

Advantages

1. A simple technique for the separation of the components.

2. In this procedure, fewer kinds of equipment are utilized. The separation is done in an extremely brief time frame as the components elute quickly.

3. The non-volatile mixtures can be isolated by this technique.

Disadvantages

Since the evaporation rate depends on the boiling point and the solvents like hexane will evaporate faster as compared to others. As a result, the mobile phase becomes more polar.

Did you know?

Thin layer Chromatography (TLC) was developed in 1983 by Izmailov and Shraiber, based on Mikhail Tswett's earlier description of column chromatography.

Gas Chromatography

Aim

Determination of lipids from subcutaneous fat by gas chromatography.

Principle

The component in the mixture is appropriated between two phases, one of which is the stationary phase and the other is a mobile phase gas, or transporter gas, that helps the combination through the stationary phase.

Mixtures in the mobile phase interact with the stationary phase as they go through because of the distinctions in properties and structure of each component, the size and affinity of each interaction with the stationary phase are unique. Therefore, under the same driving force, the retention time of different components differs in the column hence moving out of the column in various orders [1 - 4].

Application

1. In the food industry for ensuring the safety of the food items.

2. Used for analysis of sterols, oils, low chain fatty acid, aroma components and off-flavours, and many contaminants, such as pesticides, industrial pollutants and certain types of drugs in food.

Advantages

1. High-resolution power is needed as compared to other methods.

2. This method has high sensitivity.

3. It has high accuracy and precision.

4. Analysis of the sample is carried out very quickly (minutes, even seconds).

Disadvantages

1. The limitation of GC is that the substance must be volatile so that a finite fraction of it is distributed in the gaseous phase.

2. For organic substances, volatility is rarely adequate if the molecular weight of the compound exceeds 500.

3. High temperatures of up to 300°C enhance volatility, but decomposition of the substance can be the result.

Did you know?

1. Mikhail Tsvet is regularly viewed as the "Father of chromatography". He dealt with isolating various chlorophylls, which are exceptionally colourful. Both test and chroma mean colour.

2. You could likewise give a shoutout to Erika Cremer, who is another of these overlooked scientific experts. She was entirely influential and first proposed gas chromatography.

High-performance Liquid Chromatography of Lipids

Refer to Chapter 17.

Spectroscopic Analysis

Infrared Spectroscopy

Aim

Lipid quantification method using Infrared spectroscopy.

Principle

Interestingly, polarized FTIR spectroscopy is carried out on lipid monolayers or multilayers that stack to provide information on the orientation of the various chemical bonds for their membrane plane [1 - 5].

Applications

1. FTIR spectroscopy can be straightforwardly applied to microorganisms in suspensions.

2. It permits quick and affordable detection of the principal parts of natural

material (lipids, proteins, nucleic acids and sugars) by their particular absorbance frequencies.

Advantages

High Scan Speed

Infrared spectroscopy can get data for the entire range of frequency simultaneously, within one second. In this way, IR can be utilized to investigate a substance that isn't entirely steady and finish the scan before it begins to deteriorate.

High Resolution

The resolution of the overall prism spectrometer is just around 3 cm-1, but the resolution of the infrared spectrometer is a lot higher. For instance, the resolution of the Grating infrared spectrometer could be 0.2 cm^{-1}, and the goal of the FT infrared spectrometer could be 0.1-0.005 cm^{-1}.

Disadvantages

Sample Constraint

Infrared spectroscopy isn't appropriate for the sample that contains water since this solvent ingests IR light.

Quantification

Infrared spectroscopy functions admirably for the qualitative analysis of a large variety of samples, yet quantitative analysis might be restricted under specific conditions like high and low focuses.

Did you know?

1. In firefighting, infrared cameras are being utilized to find individuals or animals in heavy spots and to identify hot spots in forest fires.

2. Snakes and fish are two creatures that utilize infrared light to catch their prey and protect themselves.

Nuclear Magnetic Resonance Spectroscopy

Aim

Quantification of liver fat using magnetic resonance spectroscopy.

Principle

Magnetic resonance imaging (MRI) using standard pulse sequences is insensitive to diffuse fatty infiltrations. The signals from fat and water can be discriminated in magnetic resonance (MR) experiments, due to a 3-4 ppm difference in resonance frequency. MR spectroscopy, therefore, provides a unique opportunity for the noninvasive quantification of fat [1 - 4].

Applications

1. NMR (Nuclear Magnetic Resonance) is utilized in science to study biofluids, cells, perfused organs and biomacromolecules like nucleic acids (DNA, RNA), carbohydrates, proteins and peptides.

2. NMR is utilized in physical chemistry to study high-pressure diffusion, liquid crystals, liquid crystal solutions, membranes, and rigid solids.

3. NMR is utilized in food science.

Advantages

1. With a computer assembly, we can figure out the entire 3D construction of proteins and enzymes.

2. The movement of segments (domains) can be analyzed.

3. This technique can lead us to the perception of chemical energy.

4. We can research the dielectric constant, the polarity and some other properties of the solvent or some additional material.

Disadvantages

1. This is useful for more precise assurance of the construction, yet not for the accessibility of higher atomic weight.

2. The resolving power of NMR is less than some different kinds of investigations (for example, X-Ray crystallography) since the data acquired from a similar material is significantly more complex.

3. The expense of the trial execution is expanding with the higher strength and the complexity of determination.

Did you know?

Swiss researcher Richard Robert Ernst was granted the 1991 Nobel Prize in chemistry for contributions to the advancement of the strategy for high-resolution nuclear magnetic resonance (NMR) spectroscopy.

Raman Spectroscopy

Aim

Identification and imaging of lipids dependent on confocal Raman micro-spectroscopy.

Principle

Confocal Raman micro-spectroscopy is utilized to identify lipids in microalgae rapidly and non-destructively. Microalgae cells are cultured under nitrogen deficiency. The accumulation of lipids in Scenedesmus obliquus is observed by Nile red staining, and the total amount of lipids accumulated in the cells was estimated by the gravimetric method. The signals from various microalgae cells are collected by confocal Raman micro-spectroscopy to establish a prediction model of intracellular lipid content, and surface scanning signals for drawing pseudo-colour images of lipids distribution [1 - 4].

Application

1. Regardless of whether the goal is qualitative or quantitative information, Raman's investigation can give key data effectively and rapidly.

2. It can be utilized to quickly describe the chemical composition and construction of a sample, regardless of whether strong, fluid, gas, gel, slurry or powder.

Advantages

1. Numerous organic and inorganic materials are reasonable for Raman analysis. These can be solids, fluids, polymers or vapour.

2. No sample preparation is required.

3. Not interfered with by water.

4. Non-destructive.

Disadvantages

1. It cannot be utilized for metals or composites.

2. The Raman impact is extremely feeble. The discovery needs delicate and highly optimized instrumentation.

3. The fluorescence of impurities or the actual sample can hide the Raman spectrum. A few mixtures fluoresce when illuminated by the laser bar.

4. Sample heating through exceptional laser radiation can annihilate the sample or cover the Raman spectrum.

Did you know?

1. Raman effect was given by Dr C.V. Raman in 1928, for which he got Nobel Prize in 1930.

2. In the late 1970s, a microscopic Raman with an optical microscope instrument outfitted with a Raman spectrometer appeared, and it was utilized in many fields as a local analysis device.

Fluorescence Spectroscopy

Aim

A simple and rapid protocol for measuring neutral lipids using fluorescence spectroscopy.

Principle

A basic technique for estimating the oil content of algal cells by staining culture with Nile Red in the presence of ethanol, a mild solvent. The fluorophore Nile RedStain (9-diethylamino-5H-benzo enoxacin-5-one) has been utilized to determine the presence of neutral lipids in microalgal cells. Cellular fluorescence of stained cells and gravimetric or chromatographic determinationof lipids are linearly correlated when Nile Red and excitation range is between 488 - 525 nm and the fluorescent emission range is between 570-600 nm. Nile Red is a vital stain which allows flowing cytometric sorting of live microalgal populations based on their lipid content [1 - 4].

Application

1. It is a generally acknowledged and powerful method that is utilized for a variety of environmental, industrial, clinical diagnostics, DNA sequencing,

forensics, genetic analysis, and biotechnology applications.

2. It is an important analytical device for both quantitative and qualitative analysis.

Advantages

1. Fluorescence microscopy is the most well-known technique for contemplating the unique conduct shown in live-cell imaging.

2. This stems from its capacity to isolate individual proteins with a high level of specificity amidst non-fluorescing material.

3. The affectability is sufficiently high to recognize as few as 50 atoms for every cubic micro-meter.

Disadvantages

1. Fluorophores lose their capacity to fluoresce as they are enlightened in an interaction called photobleaching. Photobleaching happens as the fluorescent particles accumulate compound harm from the electrons excited during fluorescence.

2. Cells are susceptible to phototoxicity, especially with short-frequency light. Besides, fluorescent particles tend to create responsive synthetic species when under illumination, which improves the phototoxic effect.

Did you know?

Fluorescence was first found in 1845 by Fredrick W. Herschel. He found that UV light can excite a quinine solution (for example, carbonated water) to produce blue light.

Colourimetric Analysis

Aim

Estimation of total lipids by the colourimetric method.

Principle

Unsaturated lipids react with sulphuric acids to form carbonium particles. In a subsequent advance, the carbonium particles react with phospho vanillin to give a pink-coloured complex (Fig. **16.7**). The intensity of the colour formed is directly relative to the total lipid concentration present in the sample [1 - 4].

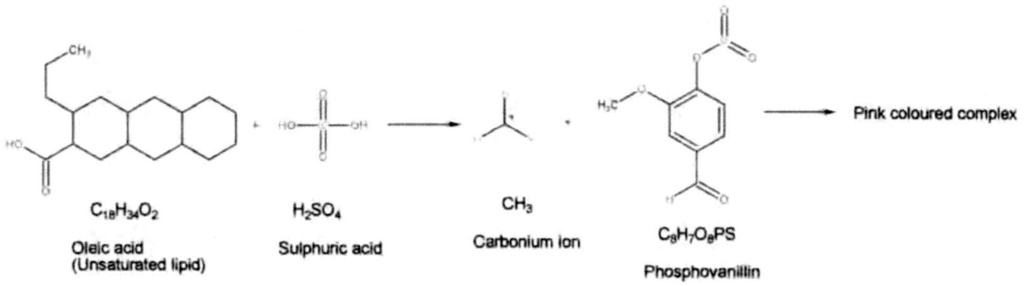

Fig. (16.7). Formation of pink colour complex.

Requirements

❖ ***Reagents***

- Phosphovainilline - 235 mmol/L
- Phosphoric acid - 2 mmol/L
- Preparation of Reagent: Add 350 ml of the vanillin reagent and 50 ml of water to a flask. Add, with consistent mixing, 600 ml of concentrated (85%) phosphoric acid. This solution is likewise stable for around 2 months in a brown bottle at room temperature.
- Standard - Total lipids aqueous primary standard 750 mg/dL
- Conc. Sulfuric Acid

❖ ***Plastic wares***

- Pipettes
- Test vials or cuvettes

❖ ***Miscellaneous***

- Timer
- 100°C heating bath
- Control serum
- Spectrophotometer

❖ ***Specimen collection*** Serum or Plasma

Storage and Stability

1. All reagents are stable until the expiration date shown on the label. Total lipid Reagent stored at 2-8°C.

2. Protected from light.

3. Stability of the example: Total lipids are stable for 24 h at room temperature (15-25ºC) or 3 days at 2-8ºC.

4. Signs of reagent deterioration: o Presence of particles and turbidity. o Blank absorbance (A) at 520 nm ≥0.32.

Procedure

1. Assay conditions: Wavelength............................ 520 nm (490-550).

Cuvette: ....1 cm light path.

Temperature ....37°C.

2. Adjust the instrument to zero with distilled water.

3. Pipette into a glass tube: (Table **16.3**)

Table 16.3. Procedure for estimation of total lipids by the colourimetric method.

-	Standard	Sample
H_2SO_4 (mL)	2.5	2.5
Standard (μL)	100	-
Sample (μL)	-	100

4. Mix and incubate for 10 minutes at 100°C in a boiling water bath.

5. Cool in iced water and transfer from Table 1 into a cuvette in Table **16.4**.

Table 16.4. Procedure for estimation of total lipids by the colourimetric method (digestion).

-	Blank	Standard	Sample
Reagent (ml)	1.0	1.0	1.0
Sample Acid Digest (μL) (Solution B)	-	-	50
Standard Acid Digest (μL) (Solution A)	-	50	-

6. Mix and incubate for exactly 15 minutes at 37°C.

7. Read the absorbance (A) of the samples and standard, against the Blank. The colour is stable for at least 1 hour.

Calculation

$$Total\ Lipid\ in\ sample\ (mg/dl) = \frac{Absorbance\ -\ Blank\ X\ 750\ (Conc. of\ Standard)}{Standard\ -\ Blank}$$

Normal Value

Serum or plasma: 450-800 mg /Dl.

Results

Results obtained using ATLAS reagents (y) did not show systematic differences when compared with other commercial reagents (x). The results of the performance characteristics depend on the analyzer used.

Precautions

1. Exercise the typical safety measures needed for the treatment of all lab reagents.

2. Pipetting by mouth isn't suggested for any lab reagent. Concentrated sulfuric corrosive cause serious burns and eye harm.

Troubleshooting

1. Total lipids cal: Proceed cautiously with this product because, due to its tendency, it can get contaminated without any problem.

2. Calibration with the aqueous standard may cause a systematic error in programmed procedures. In these cases, it is prescribed to utilize a serum Calibrator.

3. Use clean expandable pipette tips for its dispensation.

Application

This colourimetric assay gives the accompanying benefits over different techniques for lipid estimation:

1. Background absorbance can be effectively rectified for each well.

2. There is less danger of dealing with and transferring sulfuric acid contained in reaction combinations.

3. Colour develops more reliably, giving a more exact estimation of absorbance.

4. The assay can be utilized for quantitative estimation of lipids extracted from a wide assortment of sources.

Advantages

1. It is quick, economical, and has a simple activity of a spectrometer.

2. It is a very quick and convenient strategy to that of the gravimetric or volumetric cycles, and they are effortlessly upgraded for computerization.

3. Deal with the Colourimeter, doesn't need an experienced individual.

Disadvantages

1. The primary bottleneck of this strategy is that colourless mixtures can't be analyzed.

2. It needs a huge measure of samples for investigation.

3. It has low sensitivity.

4. Similar shades of interfering material can make errors in outcomes.

Did you know?

The Duboscq colourimeter was created by Jules Duboscq in 1870. The Duboscq colourimeter was the most widely manufactured and utilized of the different colourimeter designs.

CONCLUSION

Lipids are an essential dietary component for humans. Due to various applications in living organisms, it is essential to understand the type of lipids qualitatively and quantitatively as well.

VIVA VOCE

1. What is the formula used for determining the total lipid content?

Ans:

$$Total\ Lipid\ (g) = \frac{Weight\ of\ lipid\ in\ aliquot\ X\ Volume\ of\ chloroform\ layer}{Volume\ of\ aliquot}$$

2. What are the advantages of gravimetric analysis?

Ans:

1. Used to determinc thc atomic masscs of various elements.

2. Provides negligible instrumental error and does not require a series of standards for calculations of unknown samples.

3. Does not require expensive equipment.

3. What are the limitations of this experiment?

Ans:

- Usually gives an analysis of a single element, or a limited group of elements, at a time.

- Relatively time-consuming.

4. What is acid value?

Ans: The acid value is characterized as the no. of milligrams of potassium hydroxide required to neutralize the free fatty acids present in one gram of fat.

5. What is the formula used for the determination of acid value?

Ans:

$$Acid\ value = \frac{56.1\ x\ V\ x\ N}{W}$$

Where, V = Volume in mL of standard potassium hydroxide or sodium hydroxide used.

N= Normality of the potassium hydroxide solution or sodium hydroxide solution.

W= Weight in gm of sample.

6. What is saponification?

Ans: Saponification is the hydrolysis of fats or oils under fundamental conditions to manage the cost of glycerol and the salt of the corresponding fatty acid.

7. Which indicator is used in a saponification value in an oil experiment?

Ans: Phenolphthalein indicator.

8. What is the formula used to determine saponification value?

Ans:

$$Saponification\ value = \frac{Volume\ of\ acid\ required\ to\ neutralize\ KOH\ X\ Equivalent\ factor\ x\ 1000}{W}$$

Where, W= Weight in gm of sample

Where, W= Weight in gm of sample

9. Which colour is developed when iodine reacts with the starch?

Ans: Purple colour (End-point).

10. What is the normal range of blood cholesterol levels?

Ans: 130 - 250 mg/dL.

11. What are the two applications of TLC?

Ans:

1. Separation of different lipid classes from animal and plant tissues.

2. Determine the number of components in a mixture to determine the identity of two substances or monitor the progress of a reaction.

12. What is the principle of TLC?

Ans: The adsorption-desorption distribution coefficient (Kd) is a significant parameter for understanding the mobility of a compound in the environment (partitioning) and its distribution between water, sludge, soil and sediment compartments.

14. What is the principle of gas chromatography?

Ans: The sample solution injected into the instrument enters a gas stream which transports the sample into a separation tube known as the "column." (Helium or nitrogen is used as the so-called carrier gas.) The various components are separated inside the column.

15. What is the principle of HPLC?

Ans: The separation principle of HPLC depends on the circulation of the analyte (sample) between a mobile phase (eluent) and a stationary phase (packing material of the column). Depending upon the chemical structure of the analyte, the particles are retarded while passing the stationary phase.

16. What is the principle of IR spectroscopy?

Ans: IR spectroscopy recognizes the retention of light by a compound, in the IR region of the electromagnetic spectrum. To absorb light a particle should include a bond inside its structure that can display what is alluded to as a 'dipole moment' which implies electrons inside a bond are not shared similarly.

17. What are the applications of IR spectroscopy?

Ans:

- FTIR spectroscopy can be straightforwardly applied to microorganisms in suspensions.

- It permits quick and affordable detection of the principal parts of natural material (lipids, proteins, nucleic acids and sugars) by their particular absorbance frequencies.

18. What are the advantages of IR?

Ans: High resolution: The resolution of the overall prism spectrometer is just around 3 cm-1, but the resolution of the infrared spectrometer is a lot higher. For instance, the resolution of the Grating infrared spectrometer could be 0.2 cm^{-1}, and the goal of the FT infrared spectrometer could be 0.1-0.005 cm^{-1}.

19. What is the approach of this experiment?

Ans: Excite the nuclei sample into nuclear magnetic resonance with the help of radio waves to produce NMR signals. These NMR signals are detected with sensitive radio receivers. The resonance frequency of an atom in a molecule is changed by the intramolecular magnetic field surrounding it.

20. What is meant by liver biopsy?

Ans: A liver biopsy is a procedure to remove a small piece of liver tissue, so it can be examined under a microscope for signs of damage or disease. Your doctor may recommend a liver biopsy if blood tests or imaging studies suggest you might have a liver problem.

21. What are the applications of NMR spectroscopy?

Ans:

- NMR (Nuclear Magnetic Resonance) is utilized in science to study biofluids, cells, perfused organs and biomacromolecules like nucleic acids (DNA, RNA), carbohydrates, Proteins and peptides.

- NMR is utilized in physical and physical chemistry to study high-pressure diffusion, liquid crystals, liquid crystal solutions, membranes, and rigid solids.

- NMR is utilized in food science.

22. What do you mean by data mapping?

Ans: Data mapping is the way toward coordinating with fields starting with one data set and then onto the next. It's the initial step to working with data migration, data integration, and other data management tasks.

23. Which stain is used in the Raman spectroscopy experiment and why?

Ans: In this study, Nile Red, a fluorescent dye that has been utilized to distinguish the presence of lipid bodies in various types of microorganisms, is fused into a simple, quick, and reliable protocol for estimating the neutral lipid content of *Auxenochlorella protothecoides*, a green alga.

24. The emission peak for neutral lipids is observed at which wavelength?

Ans: 590 nm.

25. Secondary peaks representing more polar lipids can be observed at what wavelength?

Ans: 645 nm.

26. What is the REFERENCES value or normal value of serum/plasma?

Ans: Serum or Plasma: 450-800 mg /Dl.

REFERENCES

[1] Meullemiestre, A; Breil, C; Abert-Vian, M; Chemat, F. *Modern Techniques and Solvents for the Extraction of Microbial oils*; Springer Briefs in Molecular Science, **2015**. [http://dx.doi.org/10.1007/978-3-319-22717-7]

[2] Saenge, C.; Cheirsilp, B.; Suksaroge, T.T.; Bourtoom, T. Potential use of oleaginous red yeast Rhodotorula glutinis for the bioconversion of crude glycerol from biodiesel plant to lipids and carotenoids. *Process Biochem.,* **2011**, *46*(1), 210-218. [http://dx.doi.org/10.1016/j.procbio.2010.08.009]

[3] Nelson, D.L.; Lehninger, A.L.; Cox, M.M. *Lehninger principles of biochemistry*; Macmillan, **2008**.

[4] Yates, D.W. An introduction to practical biochemistry. McGraw-Hill Book Company (UK) Ltd: London, **1978**; p. 362.

[5] Meek, D.R.; Mills, P.R.; Gray, H.W.; Duncan, J.G.; Russell, R.I.; McKillop, J.H. A comparison of computed tomography, ultrasound and scintigraphy in the diagnosis of alcoholic liver disease. *Br. J. Radiol.,* **1984**, *57*(673), 23-27. [http://dx.doi.org/10.1259/0007-1285-57-673-23] [PMID: 6704644]

CHAPTER 17

Commonly Used Basic and Advanced Techniques

Anshika Sah[1], **Sinchan Hait**[2] and **Sai Joshi**[3,*]

[1] *Environmental Studies and Resource Management, TERI School of Advanced Studies, New Delhi - 110 070, India*

[2] *Department of Chemical Engineering, BITS-Pilani K.K. Birla Goa Campus, Goa-403726, India*

[3] *CSIR-NCL, Dr. Homi Bhabha Road, Pune - 411 008, India*

Abstract: Different techniques are used in a biochemical laboratory to measure various parameters. For example, a pH meter is used to measure the pH of the sample and electrochemical cells are used to measure the redox potential. Zeta potential is another property which can be measured using zeta cells. Chromatography is a technique used to separate the components of a mixture. Two of the most common chromatography techniques are HPLC and ion exchange chromatography. HPLC is based on the interactions of the analytes in stationary as well as mobile phase. Whereas ion exchange chromatography is based on the interaction of ions and charged site of the stationary phase.

Keywords: Electrochemical cell, Electron transfer, HPLC, Ion exchange, pH, pH meter, Redox, UV detection, Zeta potential, Zeta cell.

INTRODUCTION

Measurement of certain properties of the sample is important for biochemical analysis and chemical interpretations. Redox potential is the tendency to lose and gain electrons. This can be measured by using an apparatus using electrochemical cells. Measuring pH is used to measure the acidity or basicity of a sample which can be achieved by using a pH meter. Similarly, zeta potential is an important property of colloidal solutions and is measured using zeta cells. Chromatography is a technique used to separate the components of a mixture. This involves a mobile phase and a stationary phase. The two-chromatography discussed is High-Performance Liquid Chromatography (HPLC) and ion exchange chromatography.

* **Corresponding author Sai Joshi:** CSIR-NCL, Dr. Homi Bhabha Road, Pune - 411 008, India; E-mail: saijoshi1997@gmail.com

Pamela Jha (Ed.)

MEASUREMENT OF REDOX POTENTIAL

Aim

To measure the redox potential of a sample using electrochemical cells.

Principle

The tendency of chemical species to acquire and lose electrons is called redox potential.

In an electrochemical cell, the redox reaction is not direct. The oxidation and reduction occur in different vessels, and the electrons are transferred from the reducing agent to the oxidizing agent through connecting wires. Hence in such cells, chemical energy is converted into electrical energy [1].

Half-cells are created by placing a piece of metal into a solution comprising a cation of the metal (*e.g.*, Cu metal in a solution of $CuSO_4$ or Cu^{2+}). A porous barrier or a salt bridge separates the two half-reactions. With a voltmeter, the (+) lead will make contact with one metal and the (–) lead with another (Fig. **17.1**). If a positive voltage is shown on the screen, the cell is correctly connected. The metal attached to the (+) lead is the cathode (reduction) and thus has a higher, more positive reduction potential. The metal attached to the (–) lead is the anode (oxidation) and has a lower, more negative, reduction potential.

Requirement

1. Millivolt
2. Redox electrodes
    a. Platinum and reference electrodes
    b. Combination electrodes
3. Thermometer (liquid in a glass or thermistor type)
4. Filter paper
5. Nitrate solutions of Cu, Zn (*i.e.* $Cu(NO_3)_2$, $Zn(NO_3)_2$)
6. 1 M $NaNO_3$ solution
7. Well plate
8. Alligator clips

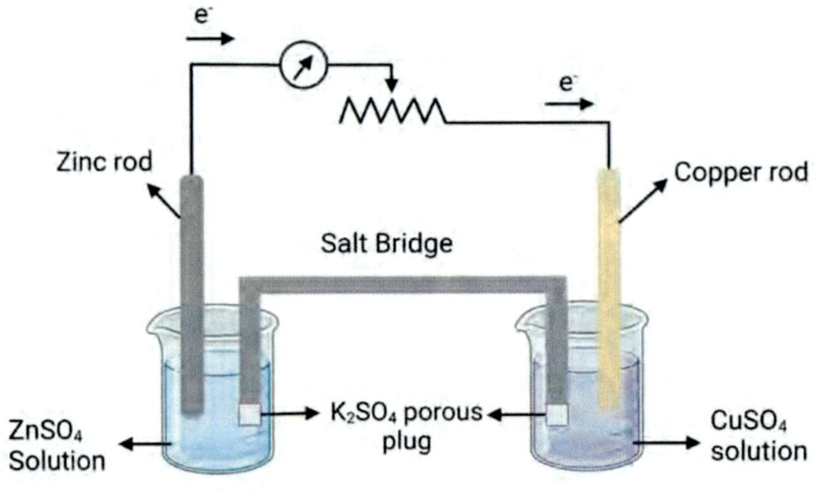

Fig. (17.1). Zn-$CuSO_4$ reaction in an electrochemical cell.

Procedure

1. Obtain a piece of filter paper and cut it into strips to use as salt bridges.
2. Add 1-2 mL of each metal nitrate $Cu(NO_3)_2$, and $Zn(NO_3)_2$ solutions to a different well in a well plate.
3. Add several drops of 1 M $NaNO_3$ to each filter paper strip. Place one end in the well containing the copper solution and the other end in the well with the zinc solution.
4. Clip metal copper into one alligator clip and zinc in the other. Place the metals in their respective solutions. If the voltage displayed in the meter is negative, then reverse the leads.
5. With a positive voltage displayed, wait about five seconds to take a voltage reading, and record the value. Be sure that you correctly label your data as you collect it.

Results

Theoretical values for these cells are:

[Zn(s) | Zn_2+(aq)] and [Cu_2+(aq) | Cu(s)] = 1.10 V

[Zn(s) | Zn_2+(aq)] and [Fe_2+(aq) | Fe(s)] = 0.32 V

[Fe(s) | Fe_2+(aq)] and [Cu_2+(aq) | Cu(s)] = 0.78 V

[Zn(s) | Zn_2+(aq)] and [Ag+(aq) | Ag(s)] = 1.56 V*

[Cu(s) | Cu_2+(aq)] and [Ag+(aq) | Ag(s)] = 0.46 V*

*This assumes that the solution of silver ions is 1.0 mol dm^{-3}.

Based on these theoretical values, the value for the zinc half-cell is Eɵ [Zn_2+(aq) | Zn(s)] = −0.76 V.

Precaution

1. The salt solutions of aqueous metal can irritate in case of contact with the skin. Thoroughly wash the affected area with water for 15 minutes.
2. Rinse the well plate using distilled water and dry it properly before using.

Advantages

1. The measurement of redox potential enables quick and relatively easy characterization of the degree of reduction of a chemical reaction.
2. Redox potentials also have applications in predicting the stability of the various compounds that play an important role in the regulation of nutrients and the availability of metals in sediment and soils.

Disadvantages

1. Contamination of the electrode surface, salt bridge, or internal electrolyte can be caused by organic matter and sulfide. It can cause drift or erratic performance when reference electrodes are used.

Troubleshooting

1. Voltage measurements can be incorrect due to the production of coating on the platinum electrode caused by hydrogen sulfide, which interferes with the measurement. This can be avoided by making sure the electrode is not inside the sulfide-rich water for too long.

MEASUREMENT OF pH

Aim: Measurement of pH by pH meter.

Principle

An electrode system that is sensitive to changes in the H+ ion concentration of a solution is used for pH measurement. The electrode system consists of a sequence of electrodes whose potential rises with pH (H+ concentration of the solution) [1] (Fig. **17.2**).

$$\mathbf{pH = pKa + \log [A^-] / [HA]}$$

Requirements

1. pH meter
2. Standard pH solutions (4, 7, 9.2)
3. Distilled water
4. 0.1M HCl
5. 0.1M NaOH
6. Vinegar

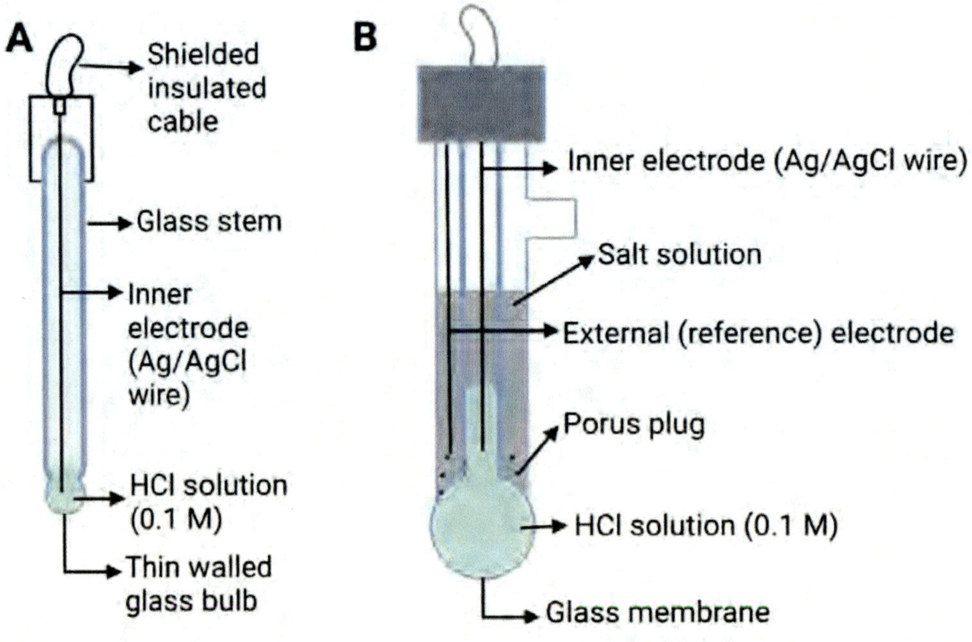

Fig. (17.2). Common pH electrodes: (**A**) glass electrode; (**B**) combination electrode.

Procedure

Calibration

1. Turn on the device.
2. Set the meter in the pH mode.
3. Calibrate the device by pressing the "CAL" button.
4. Rinse the electrode and place it into the 4.0 buffer.
5. Again, calibrate the device.
6. Rinse the electrode and place it in the buffer of pH 7.0. Wait for a stable pH reading.
7. The meter is now ready to take pH measurements.

pH Measurement

1. Take some distilled water in a clean beaker.
2. Take the electrode out, wash it with distilled water and wipe it clean with tissue paper.
3. Dip the electrode in the water.
4. Note the reading.
5. Repeat the steps for 0.1M HCl solution, 1M NaOH, and vinegar.

Results

S. No	Sample	pH
1	D.W Water	7
2	HCl	1
3	NaOH	14
4	Vinegar	2

Advantages

- As pH meters use a readable digital display, they provide a much more precise result than pH papers and titration methods.
- pH meters are portable and provide quality precision readings.

Disadvantages

- In comparison to pH paper, the maintenance of a pH meter is difficult as it requires regular cleaning to avoid contamination of samples.
- The probe has a glass tip which is highly fragile and can be damaged easily on exposure to a corrosive substance.

Troubleshooting

- If the pH values are drifting, and the response time is slow, it should be ensured that the bulb should be submerged in the pH stock solution that is 1M KCl.

Precautions

1. The tip of the probe is fragile.
2. The volume of 1 M KCl in the reference electrode needs to be constantly maintained.
3. The temperature knob should be properly immersed in the solution.

MEASUREMENT OF ZETA POTENTIAL

Aim: Measuring the zeta potential of the samples using zeta cells.

Principle

Zeta potential is a physical property exhibited by all solid-liquid and liquid-liquid colloidal systems. It is the strength of a particle's shear plane during electrophoresis. The mobility of the material is related to the dielectric constant, the viscosity of the suspended liquid, and the electrical potential at the slipping plane. This plane is the boundary between the moving material and the liquid; the point where the Stern layer meets the diffuse layer. The Stern layer is rigidly bound to the colloid but not to the diffuse layer. As a consequence, the electrical potential at this junction is related to particle mobility and is called zeta potential (Fig. **17.3**) [2 - 4]

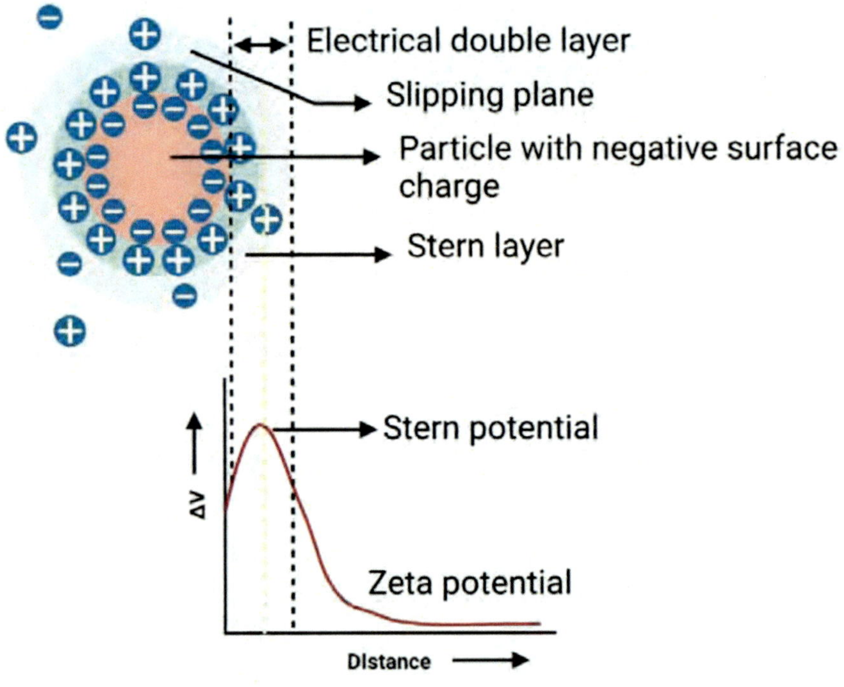

Fig. (17.3). Zeta potential of a negatively charged particle.

Measurement of Zeta Potential

Zeta potential is measured by applying an electric charge across the sample in the zeta cell.

The particles with high zeta potential get attracted to the electrode of the opposite charge.

The velocity of the particle will be proportional to the degree of zeta potential and is measured by Doppler anemometry, which analyzes speed and direction.

Requirements

1. The instrument for measuring laser Doppler velocimetry – phase analysis light scattering (LDV-PALS) or zeta potential
2. Zeta cells
3. Caps for zeta cells
4. Zeta potential transfer standard
5. 10 mM NaCl
6. Analyte nanoparticle solution
7. PAMAM (Poly(amidoamine))

Procedure

1. Rinse the zeta cells first with distilled water, followed by ethanol and again with water.
2. Dry the zeta cells with the help of nitrogen to remove the remaining solvent.
3. Prepare the samples (nanoparticles of PAMAM and gold) in a low ionic strength medium. 10 mM NaCl is advised. Suspending medium should be filtered before sample preparation using a 0.1 mm or smaller pore-size membrane.
4. Sample concentration is particle-dependent. The ideal sample concentration is decided by the light-scattering properties of the nanoparticles under analysis.
5. Transfer the sample to a 1-mL disposable syringe, remove air bubbles, inject the syringe and fill one of the ports on the zeta cell.
6. Cap both ports simultaneously to ensure even sample levels on both sides of the zeta cell.
7. The weld line of the zeta cell should be facing the front of the instrument.
8. Perform a minimum of three runs per sample to establish measurement repeatability.

Advantages

Zeta can be used to predict the long-term stability of particles. For example, particles with zeta potentials larger than ±60 mV have excellent stability, whereas particles with zeta values between -10 mV and +10 mV will experience rapid agglomeration unless they are sterically protected.

Zeta potential measurements are used in the following processes:

1. Preparation of colloidal dispersions for pharmaceutical and cosmetic preparations, food products, dyestuffs, foams, and agricultural chemicals.
2. Use of colloidal dispersions in the manufacturing processes in ceramic casting, cement and plaster, brickmaking, and pottery.
3. Disruption of undesired colloidal dispersions in water purification, the fining of wines and beers, sewage disposal, the dewatering of sludges, and the dispersal of aerosols and fogs.

Disadvantages

1. Several factors influence the magnitude and even the sign of the ZP: nanoparticle properties, aggregation, solvent conditions, electrode blackening, and unintentional development of organic macromolecules (*e.g*., proteins).

Troubleshooting

1. Sharp spikes can be observed due to the presence of dust. Filter the samples to improve measurements and make sure that zeta cells are capped until used to prevent contamination due to dust.

Precautions

1. Always wear appropriate personal protective equipment and take appropriate precautions when handling nanomaterial.
2. Follow your facility's recommended disposal procedure for your specific nanomaterial.
3. Zeta potential is dependent on pH, and therefore the sample pH should be measured before and after the zeta potential readings.
4. Check for any air bubbles in the zeta cell or on the electrodes; gently tap the zeta cell to dislodge them.
5. Before loading the zeta cell into the instrument, carefully wipe the measuring windows with lens paper.

SEPARATION USING HPLC

Aim: Separation of a sample using the HPLC.

Principle

The principle of separation of HPLC or any LC (liquid chromatography) is based on the distribution of mixture/sample(analyte) between the mobile phase(eluent) which moves in a definite direction and the stationary phase (packing material in the column) (Fig. **17.4**). The analytes have different types of interactions with both phases. The solutes that have stronger interactions with the stationary phase ("the fixed phase") tend to move slower than others with lower or no interactions with the stationary phase. So, different constituents of the sample are eluted at different times and chromatographic separation is achieved.

Consequently, the constituents eluted at different times from the column reach the detector and are monitored by it over a chromatographic run. The results obtained are shown in chromatograms [5].

Quercetin, a flavonol, is a plant-derived flavonoid that is used as a nutritional supplement. A reversed-phase HPLC method with UV detection can be used for the determination of quercetin in the extracts of *Psidium guajava, Vitis vinifera,* and extracts rich in quercetin and other flavonols in the flavonoid family [6, 7].

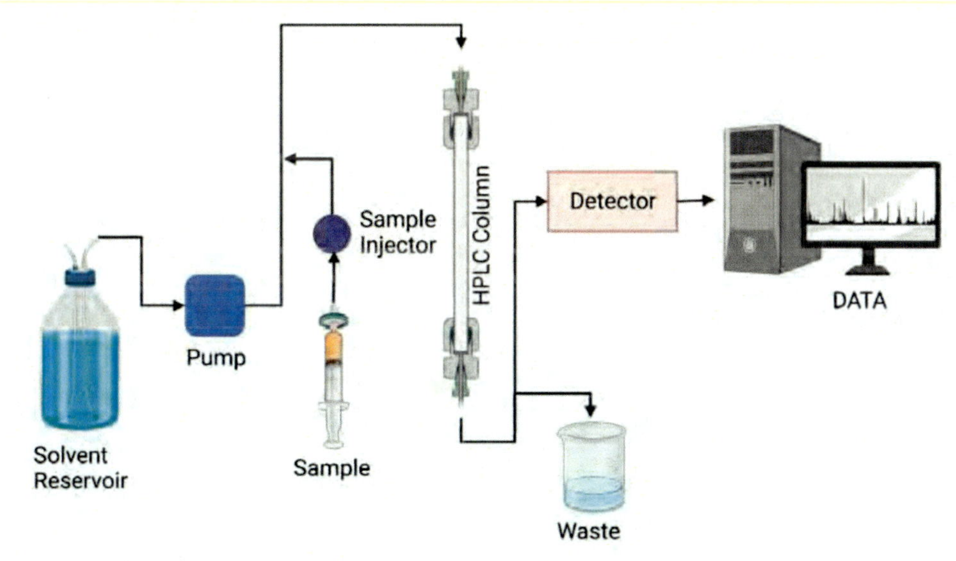

Fig. (17.4). HPLC System.

Requirements

1. Standard quercetin
2. Sample 1 - Plant leaves powder (Dried *Sedum sarmentosum)*
3. Sample 2 - Milk powder
4. Methanol
5. Trifluoroacetic acid
6. 25% hydrochloric acid solution
7. Aqueous trichloroacetic acid (1%, v/v)
8. Millipore filter

Instruments

1. HPLC with UV detector:

An HPLC is equipped with the following systems:

Solvent Delivery System

It consists of solvent reservoirs (usually glass or stainless steel of 200-2000ml), high-pressure pumps, a mixing chamber, and degassers.

Solvent reservoirs have solvent treatment that removes dissolved gases and dust to avoid band spreading and interference with the performance of detectors. They also have filters to remove the solid particles in the suspension of the mobile phase.

Degassers may consist of a vacuum pumping system, distillation system, device for heating and stirring, or system of sparging, in which the dissolved gases are swept out of solution by fine bubbles of an inert gas that is not soluble in the mobile phase.

The most common pumps used are either reciprocating or displacement pumps.

Reciprocating pumps have a motor-driven reciprocating piston to control the flow of the mobile phase, whereas displacement pumps are composed of a one-directional motor-driven plunger that pushes the mobile phase present in a syringe-like chamber.

Different solvent suppliers and a mixing chamber are also equipped with modern HPLC instruments nowadays.

The solvent delivery systems may operate using two different elution modes: isocratic and gradient elution.

In isocratic elution, the elution of the solvent ratio is kept constant over the chromatographic run, *i.e.*, the makeup of the mobile phase remains constant during separation. In gradient elution, the solvent ratio is modified over the run following a linear gradient or multistep linear gradient, *i.e.*, the composition of the mobile phase is changed during separation.

Sample Injection System

The sample injection system serves to introduce the liquid sample into the flow stream of the mobile phase. The main sample injection requirements maintain the characteristic of the sample and can withstand the high pressures of the system. Injection of samples into the column can be archived manually or by the auto-injection procedure.

Column

The column of an HPLC is usually smooth-bore stainless steel and heavy-walled glass tubing.

In these columns, the small dimensions and particle sizes of the packing materials are used.

The diameter of a column in analytical separation is about 2–5mm and its length is about 10–30 cm.

Mechanical stability is a crucial criterion for particles filled in the column, and due to the application of high pressure in HPLC, the primary requirement for columns is that they should be incompressible. For this, silica can be considered the best candidate and is predominantly used for this purpose. They can be used to create homogeneous column media of sufficient strength and with well-controlled particle size.

Detection System

The detector is a component that emits a response generated by the eluting sample compound and subsequently, signals are turned into peaks on the chromatogram.

However, the following several types of detectors listed in Table 3 are considered suitable for HPLC. Normally, the nature of analytes determines the type of detector to be applied in a study. Among these detectors, spectrophotometric, fluorescence, refractive index (RI), evaporative light scattering, and mass analyzers are the most common detector systems coupled to HPLC.

Procedure

Sample Reparation

Sample 1

1. Accurately weigh 2.0 g of sample dried *Sedum sarmentosum*.
2. Add 20 ml of methanol to the sample.
3. Reflux the solution twice, each for 30 minutes. Reflux means heating the solution but the vapours are condensed and return to the system of origin. This will allow no loss of solution during heating.
4. Filter the mixture into a 50 ml volumetric flask.
5. Wash the residue with methanol.
6. Add methanol to make up the volume and mix the solution properly by shaking.
7. Mix 5 ml of 25% hydrochloric acid solution in 20 ml of the solution.
8. Reflux the solution for 30 min.
9. After cooling the solution transfer, it to a 50 ml volumetric flask.
10. Make up the volume with methanol and shake well.
11. Filter through a 0.45 μm millipore filter before HPLC analysis.

Sample 2

1. Accurately weigh 2.0 g of milk powder samples into 50 mL centrifuge tubes.
2. Add 15 ml of aqueous trichloroacetic acid and 5 mL of acetonitrile to the sample.
3. Vortex shakes the solution for 1 minute.
4. Place the sample in an ultrasonic bath for 30 minutes and again shake it for 10 minutes.
5. Centrifuge the sample for 10 minutes at rpm ≥ 10,000.
6. Make up the volume with trichloroacetic acid and shake well.
7. Filter through a 0.45 μm millipore filter before HPLC analysis.

HPLC Method Development

Standard and the isolated fraction of quercetin were analyzed by HPLC technique using the following conditions:

It must be ensured that the sample is fully dissolved in the mobile phase.

1. Selection of the Stationary Phase (Column Medium)

- Unmodified silica or diol media should be used for nonpolar, water-insoluble materials.
- For amino group-containing polar, water-soluble substances, reverse-phase C18, C8, or C4-modified silica media (especially for large hydrophobic peptides and proteins) can be used.
- Smaller particle size matrices (=<3 µm) should be used for analytical purposes to increase the efficiency of separation.
- 5-10-µm particle sizes are suitable for semi-preparative and preparative purposes.
- The commonly-used 100-Å pore-size media are suitable for low molecular weight metabolites, amino acids, and small peptides, and for macromolecules, high porosity (wide pore), and 300-Å pore-size media are to be chosen.

2. Selection of the Mobile Phase

- The most important characteristics of the mobile phase include purity, viscosity, UV transparency, and miscibility with other solvents.
- HPLC techniques require "HPLC grade" solvents, including water specially purified for this purpose and the mobile phase should be optically pure.
- Most chromatographic detectors operate in the UV range, so HPLC reagents like acetonitrile and other quality reagents should be used, which enable photometric detection even at a wavelength of 200 nm. Less expensive grades of acetonitrile are sufficient to use for the range of 280-340 nm.

3. Selection of Column Size

The most commonly used type of column is the standard column with a 4.6-mm diameter.

These columns can be used for both analytical and semi-preparative purposes. The volume of standard columns is 10-50 µL in the case of analytical uses; whereas, for semi-preparative work, column volumes up to 1000 µL can be used.

Results

A series of peaks on the chromatogram will be recorded as the output. Each peak represents a compound in the mixture which passes through the detector and absorbs UV light. By comparing the retention time with the standard retention time of the compound, one can identify the presence of that compound in the given sample.

For quercetin, the retention time is around 8.5 minutes.

Troubleshooting

- The formation of bubbles when the solution leaves the column and the reduced pressure, which may severely interfere with photometric detection, can be caused by the mixing of the solvents during gradient elution. Hence it should be avoided.
- Pre-column can be used before the main column, with the two columns having identical column media. As it is cheaper than a column and can be replaced if blocking occurs.

Precautions

- Solvents with high viscosity should be generally avoided as their use increases the system pressure.
- Components of the mobile phase must be degassed and filtered before usage.

Advantages

1. This technique has a higher resolution and speed of analysis.
2. HPLC sections can be reused without repacking or regeneration.
3. More prominent reproducibility because of close control of the parameters influencing the effectiveness of separation.

Disadvantages

1. HPLC can be a costly technique. It requires an enormous number of expensive organics, needs a power supply, and customary support is required.
2. It very well may be complicated to troubleshoot issues or foster new techniques.
3. The absence of a universal detector for HPLC, nonetheless, the UV-Vis detector just distinguishes chromophoric compounds.

ION-EXCHANGE CHROMATOGRAPHY

It is based on the interaction occurring between positively-charged particles and the negatively-charged phosphates in DNA under low salt conditions. The contaminants are washed using high-salt solutions while DNA is eluted under high-salt conditions

Anion Exchange

Ion exchange chemistry is based on the interaction that occurs between positively-charged particles and the negatively-charged phosphates that are present in DNA. The DNA binds under low salt conditions, and contaminating proteins and RNA can then be washed away with higher salt solutions. The DNA is eluted under high salt conditions and then recovered by ethanol.

Principle

The most popular examples of anion exchange chromatography employ the use of anion exchange resin. It manipulates the interaction of negatively charged phosphates from the backbone of DNA with positively charged diethylaminoethyl cellulose (DEAE) groups on the surface of the resin. The specified silica beads on the anion-exchange resin have wide pores, a hydrophilic surface coating and a high charge density. The DEAE groups can be tightly coupled because of the wide surface area of the resin [8].

The advantages of the resin include its applicability under a broad range of pH (pH 6-9) and/or varying salinity (0.1M-1.6M). This also aids in the separation of DNA from RNA and other impurities. Hence, the salt concentration and pH conditions of the buffer largely dictate whether the nucleic acid is bound or eluted out from the column. Furthermore, the binding of DNA to the DEAE group occurs over a broad range of salt concentrations. This allows DNA to be bound to DEAE until its subsequent elution with a high salt buffer, at the same time, impurities like protein and RNA are washed using medium-salt buffers [8].

There are various anion exchanger materials available, briefly, they can be described as:

A. Strong positively charged
B. Weak positively charged

Based on this property, the application of selected solutions of known ionic strength for adsorption and elution of nucleic acids is largely determined. The solubility of the protein in water is favourable, hence the washing of the column is performed using a solution with known ionic strength to bind nucleic acids to anion exchange column materials. Elution of DNA manipulates this known ionic strength to elute DNA using known salt concentrations in a buffer to control pH. Ideally corresponding to the lowest ionic strength at which nucleic acids will completely elute.

Purification of Nucleic Acid

This method involves the creation of a cellular lysate containing nucleic acid. Initially, the lysate is bound to the solid support matrix. The elution of nucleic acid involves an aqueous mobile phase to interrupt the binding with a higher salt concentration. Finally, the eluted nucleic acid undergoes desalting such that it is suitable for downstream applications.

Anion exchange chromatography uses weak anion exchange ligands such as DEAE coupled to silica or polymer-based matrices and strong ligands such as quaternary amines coupled to polymeric matrices.

Requirements

Pump

It provides a constant flow of eluent through the injector, column and detector.

Injector

Used for sample introduction. Liquid samples can be incorporated directly, however solid samples need to be dissolved in an appropriate solvent.

Column

This is where the binding takes place.

Suppressor

The suppressor reduces the background conductivity measurement of the ions being examined by lowering the background conductivity of the chemicals used to elute samples from the ion exchange column.

Detector

Usually employs electrical conductivity for measurement.

Data System

This involves a pre-programmed computing integrator.

Procedure

1. First, the column is packed with the ion exchanger.
2. Then, the sample is added, followed by the buffer.
3. The washing buffer also separates the tightly bound particles.
4. The column is washed until no protein is obtained.
5. Elution of DNA uses a known ionic strength buffer to elute DNA using known salt concentrations in a buffer to control pH.

Applications

- This is the most effective method for water purification. Complete deionization of water, (or) a non-electrolyte solution, is performed by exchanging solute cations for hydrogen ions and solute anions for hydroxyl ions. This is usually achieved by the method used for softening drinking water.
- In the analysis of products of hydrolysis of nucleic acids. In this way, information is gained about the structure of these molecules and how it relates to their biological function as carriers of hereditary information.

Did you know?

- The first electrochemical cell was invented by Luigi Galvani in 1791.
- Arnold O. Beckman developed the first commercially successful electronic pH meter, in the year 1934, while he was a member of the faculty of the California Institute of Technology.
- Zeta potential is a major factor to decide the adsorption of a nanoparticle in a cell membrane [4].
- The ALC-100 HPLC, developed by Waters Associates (present- Waters Corporation) was introduced in 1967 and is widely considered to be the first successful commercially available HPLC system.
- Cecil Tarbet (the organizer of Cecil Instruments) presented the world's first commercially accessible variable wavelength monitor for HPLC. By the late 1970s, new techniques in HPLC, including reversed-phase liquid chromatography, took into consideration further developed separation between very similar mixtures.

CONCLUSION

The chapter included a few of the basic and advanced techniques which are used in a biochemical laboratory. This included the measurement of pH, redox potential and zeta potential, and the separation of components from the mixture through techniques like HPLC and ion exchange chromatography.

VIVA VOCE

1. What is a Redox Potential?

Ans. Redox potential measures an aqueous system's capacity to either release or accept electrons from chemical reactions.

2. What is an oxidizing and reducing system?

Ans. When a system tends to accept electrons, it is an oxidizing system. When it tends to release electrons, it is a reducing system.

3. Is the pH of water affected by the rise in temperature?

Ans. Yes, the pH changes with the temperature change. With the increase in temperature, the molecular vibration increases. This results in water ionizing and the formation of hydrogen ions hence decreasing the pH.

4. Define pH.

Ans. pH is the negative logarithm of the hydrogen ion concentration.

pH = log10 [H+]

It is used to measure the acidity or basicity of a water-based solution.

5. What do you mean by pOH?

Ans. pOH is a measure of the hydroxide ion (OH-) concentration of a solution. pOH is the inverse log of the concentration of hydroxide in a solution.

P[OH] = -log10 [OH-]

6. Is there any other method to test pH?

Ans. pH paper and pH pen are the other two methods that we can use to test pH.

7. What is the pH of Blood?

Ans. The pH of blood ranges from 7.35 to 7.45.

8. What is a weld line?

Ans. The Weld line is a small notch along the top of the cell on the front of the zeta cell, which should be facing the front of the instrument.

9. What are the two modes of applied voltage?

Ans. The two modes are automatic mode and applied mode. In automatic mode, the most suitable voltage is determined by the software. But, it is suggested to run in manual mode and start at a low voltage, typically 80 V, and increase the voltage steadily for optimal results.

10. What are zeta cells?

Ans. Also known as folded capillary cells, these are polycarbonate cells with gold-plated electrodes. Examples are DTS1060C and Malvern Instruments.

11. Where is HPLC most generally used?

Ans. They are generally used in pharmaceutical industrial and analytical fields where they are used to test the products and to detect the raw ingredient used to make them. They are also used in forensics, clinical and food industries.

12. What is the difference between Liquid chromatography and HPLC?

Ans. The solvent in LC travels by the force of gravity under low pressure to pass the mobile phase through the column whereas, HPLC is an advanced LC where the solvent travels under high pressure using a pump. HPLC reduces the time of separation as compared to traditional LC.

13. What is predominantly used as a packing material in the column of an HPLC? Besides that, what other particles are used as packing material?

Ans. Silica is predominantly used as a packing material in HPLC. Porous particles like polystyrene-divinyl-benzene synthetic resin, alumina, or other ion-exchange resins are also used as packing material.

14. What are the basic parts of an HPLC?

Ans. The basic parts of HPLC are solvent reservoirs, pumps, a mixer/mixing chamber, a sample injector, a chromatographic column, a detector and a fractional collector.

REFERENCES

[1] Wilson, K.; Walker, J., Eds. *Principles and techniques of biochemistry and molecular biology*; Cambridge university press, **2010**. [http://dx.doi.org/10.1017/CBO9780511841477]

[2] Lunardi, C.N.; Gomes, A.J.; Rocha, F.S.; De Tommaso, J.; Patience, G.S. Experimental methods in chemical engineering: Zeta potential. *Can. J. Chem. Eng.,* **2021**, *99*(3), 627-639. [http://dx.doi.org/10.1002/cjce.23914]

[3] Clogston, JD; Patri, AK Zeta potential measurement. *Methods Mol. Biol.,* **2011**, *697*, 63-70. [http://dx.doi.org/10.1007/978-1-60327-198-1_6]

[4] Lunardi, C.N.; Gomes, A.J.; Rocha, F.S.; De Tommaso, J.; Patience, G.S. Experimental methods in chemical engineering: Zeta potential. *Can. J. Chem. Eng.,* **2021**, *99*(3), 627-639. [http://dx.doi.org/10.1002/cjce.23914]

[5] Lozano-Sánchez, J; Borrás-Linares, I; Sass-Kiss, A; Segura-Carretero, A. Chromatographic technique:

High-performance liquid chromatography (HPLC). In: *In Modern Techniques for Food Authentication*; Academic Press, **2018**; pp. 459-526.

[6] Yue-ling, M.; Yu-jie, C.; Ding-rong, W.; ping, C.; Ran, X. HPLC determination of quercetin in three plant drugs from genus sedum and conjecture of the best harvest time. *Pharmacogn. J.,* **2017**, *9*(6), 725-728.
[http://dx.doi.org/10.5530/pj.2017.6.114]

[7] Sanghavi, N.; Bhosale, S.D.; Malode, Y. RP-HPLC method development and validation of Quercetin isolated from the plant Tridax procumbens L. *J. Sci. Innov. Res.,* **2014**, *3*(6), 594-597.
[http://dx.doi.org/10.31254/jsir.2014.3609]

[8] Jungbauer, A.; Hahn, R. Ion-exchange chromatography. *Methods Enzymol.,* **2009**, *463*, 349-371.
[http://dx.doi.org/10.1016/S0076-6879(09)63022-6] [PMID: 19892182]

CHAPTER 18

Significant Experimental Hazard: Safety Data Sheet (SDS)

Sai Joshi[1,*], **Anshika Sah**[2] and **Sinchan Hait**[3]

[1] *CSIR-NCL, Dr. Homi Bhabha Road, Pune - 411 008, India*

[2] *Environmental Studies and Resource Management, TERI School of Advanced Studies, New Delhi - 110 070, India*

[3] *Department of Chemical Engineering, BITS-Pilani K.K. Birla Goa Campus, Goa-403726, India*

Abstract: A safety data sheet or SDS is a datasheet prepared by the manufacturer to illuminate the user about any potential hazard and preserved by the owner. SDS also provides information about the handling and working of the product. SDS is also known as a material safety data sheet (MSDS) or product safety data sheet (PSDS). It must include information about the physical, environmental health, and health hazards; safety precautions, and protective measurements; transportation and storing. Guidelines for the preparation of such SDS are laid down by the occupational safety and health administration (OSHA). This chapter briefly explains the SDS of the significant experimental hazards of those chemicals used in the previous chapters.

Keywords: Hazard, MSDS, OSHA, PSDS, SDS, Safety.

INTRODUCTION

SDS is a document that provides guidelines and information for working with a certain chemical and about storage and transportation. In 2012, the United States made the safety data sheet in a consistent manner, which is globally accepted. The SDS cannot be presumed to be the same for an identical chemical name, as they have different formulations based on the country of origin. An SDS is written in English and contains various sections (s). The section(s) are discussed briefly below [1 - 3]:

* **Corresponding author Sai Joshi:** CSIR-NCL, Dr. Homi Bhabha Road, Pune - 411 008, India; E-mail: saijoshi1997@gmail.com

Pamela Jha (Ed.)

Section 1

Section 1 is about "identification" and provides information about the product identifier, including synonyms, recommended usage, and supplier information. Also, any special recommendations are provided by the supplier.

Section 2

Section 2 is about "Hazards identification". It identifies potential chemical hazards, warnings, and descriptions of any hazards that are not classified. In case of an unknown mixture of ingredients, acute toxicity must be mentioned.

Section 3

This section identifies "Composition or information on ingredients". Section 3 deals with ingredients details, including impurities and stabilizing additives. If a trade secret requires a specific chemical identity, and percentage can be withheld.

Section 4

Section 4 or "First-Aid measures" deals with primary care that needs to be taken for the individual who has been compromised to the chemical.

Section 5

Section 5 or "Fire-Fighting measures" gives recommendations in case of fire caused by the chemical, advice on hazards caused by chemicals during fire and precautions.

Section 6

Section 6 or "Accidental release measures" provides recommendations for any accidental release of the chemical. Also, advice about containment, cleanup, and exposure prevention to people, and the environment; from small spills to large.

Section 7

Section 7 or "Handling and storage" deals with safe handling and safe storage of the chemical.

Section 8

Section 8 or "Exposure control or personal protection" provides information on exposure limits, control, and personal protective measures.

Section 9

Section 9 or "Physical and chemical properties" inks different physical and chemical properties linked to the chemical or mixer.

Section 10

Section 10 or "Stability and reactivity" provides information on reactivity (description of specific test data on that chemical), chemical stability (on the stability of that chemical, description of stabilizer or any change in physical appearance), and others (indication of any possible hazardous reaction, conditions to be avoided, incompatible materials, and any anticipated hazard).

Section 11

Section 11 or "Toxicological information" inks toxicological data that may arise due to exposure, or routes of exposures, LD50, symptoms, *etc*.

Section 12

Section 12 or "Ecological information" informs about the environmental impacts of these chemicals if released into the environment. Data on aquatic and terrestrial organisms, biodegradation, bioaccumulation potential, groundwater contamination, and any other adverse effect.

Section 13

Section 13 or "Disposal consideration" guides on disposal mechanism, recycling of the chemical or its container and handling. This section also advises referring to section 8.

Section 14

Section 14 or "Transport information" guides on the safe transportation of the chemical.

Section 15

Section 15 or "REgulatory information" indicates any safety, health, and environmental regulations or guidelines to be followed which is not mentioned earlier in the SDS document.

Section 16

Section 16 or "Other information" informs about when SDS was prepared, any revisions, any state of changes in previous versions, and any other useful information.

Sections 12-15 are non-mandatory [3].

HAZARD INFORMATION ON CHEMICALS USED IN THE EXPERIMENTS FROM SDS

Hazard information on chemicals used in the experiments from SDS are mentioned in Table **18.1**.

Table 18.1. Hazard information on chemicals used in the experiments from SDS.

Reagent Name	Effect on Human				Fire Hazard	Effect on Environment	For more Information
	Skin	Eye	Ingestion	Respiratory Tract			
Molisch Reagent	Corrosive, Irritant	Irritant	Acute toxicity	Shortness of breath	Flammable	Ethanol has a slight acute and chronic toxicity to aquatic life.	https://www.fishersci.com/store/msds?partNumber=S25754&productDescription=MOLISCH+REAGENT+100+ML&vendorId=VN00115888&countryCode=US&language=en
Concentrated sulfuric acid	Corrosive, Irritant	Serious damage/irritation	Causes severe swelling, severe damage to the delicate tissue and danger of perforation	Respiratory organs toxicity	Non-flammable (Do not use water)	Do not let it enter drains. It is toxic to aquatic invertebrates and algae.	https://www.fishersci.com/msdsproxy%3FproductName%3DA300700LB%26productDescription%3DSULFURIC%2BAC%2BACS%2B700LB%26catNo%3DA300-700LB%26vendorId%3DVN00033897%26storeId%3D10652
Fehling's Solution A	Sensitization, irritation	Irritation	May cause damage to organs through prolonged or repeated exposure if swallowed.	Irritation	Not determined	Do not let it enter drains. It is toxic to aquatic life.	https://www.fishersci.com/store/msds?partNumber=AC270790010&productDescription=FEHLING+B+SOLUTION+1LT&vendorId=VN00032119&countryCode=US&language=en
Fehling's Solution B	Corrosion	Causes damage	Causes organ damage. Diarrhoea. Nausea. Vomiting.	Irritation	-	Thermal decomposition can lead to the release of irritating gases and vapours.	https://www.fishersci.com/store/msds?partNumber=AC270790010&productDescription=FEHLING+B+SOLUTION+1LT&vendorId=VN00032119&countryCode=US&language=en
Benedict's reagent	Irritation	Serious damage/irritation	Nausea. Vomiting. Low arterial pressure. Enlargement/disease of the liver.	Coughing after inhalation	-	Harmful to aquatic life with long-lasting effects. Do not let it enter drains.	https://www.fishersci.dk/chemicalProductData_uk/wercs?itemCode=10325900&lang=EN
Selwinoff's reagent	Severe burns	Serious damage/irritation	-	Respiratory irritation	-	Very hazardous to aquatic life. Do not let it enter drains.	https://www.lobachemie.com/lab-chemical-msds/MSDS-SELIWANOFFS-REAGENT-CASNO-05681-EN.aspx
Barfoed's Reagent	Irritation	Eye irritation	May cause damage to organs	Irritation	-	Very toxic to aquatic life with long-lasting effects. Do not let it enter drains.	https://www.cdhfinechemical.com/images/product/msds/39_1191427268_BARFOEDSREAGENTMSDS.pdf
Bial's Reagent	Skin burns	Eye irritation	Harmful if swallowed	Toxic if inhaled	Highly flammable	It is very toxic to aquatic life. Do not let it enter drains.	https://www.lobachemie.com/lab-chemical-msds/MSDS-BIALS-REAGENT-SUMNER-CASNO-1927F-EN.aspx

(Table 18.1) cont.....

Reagent Name	Effect on Human				Fire Hazard	Effect on Environment	For more Information
	Skin	Eye	Ingestion	Respiratory Tract			
Phenylhydrazine hydrochloride	Hazardous, irritation, permeates into the skin, and skin sensitization	Hazardous. Eye irritation/damage	Very hazardous. Suspected of causing cancer	Acute Inhalation Toxicity	Combustible at high temperatures. Do not use water in case of fire.	Very toxic to aquatic organisms.	https://www.fishersci.com/store/msds?partNumber=AC151555000&productDescription=PHENYLHYDRAZINE+HYDROCHL+500GR&vendorId=VN00032119&countryCode=US&language=en
Acetic acid	Caustic burns/corrosion of the skin	Corrosion of the eye tissue.	In case of ingestion burns to the gastric/intestinal mucosa. Possible oesophagal perforation. Blood in vomit. Diarrhoea. Shock. Low arterial pressure. Enlargement/disease of the liver. Decreased renal function	Coughing. sore throat. Respiratory difficulties. Corrosion of the respiratory tract.	Highly flammable	Thermal decomposition can lead to a release of irritating gases and vapours like carbon monoxide (CO) and carbon dioxide (CO2)	https://www.fishersci.com/msdsproxy%3FproductName%3DAC423220025%26productDescription%3DACETIC%2BACID%252C%2BGLACIA%2B2.5L%26catNo%3DAC42322-0025%2B%26vendorId%3DVN00033901%26storeId%3D10652
Anthrone reagent	Skin corrosion/irritation	Eye damage/ irritation	Specific target organ toxicity	Acute Inhalation Toxicity	-	The products of degradation are toxic to aquatic life. Do not let it enter drains.	https://www.fishersci.com/store/msds?partNumber=AC401360250&productDescription=ANTHRONE%2C+95%25%2C+EXTRA+PUR+25GR&vendorId=VN00032119&countryCode=US&language=en
DNSA reagent		Serious damage/irritation	Acute oral toxicity.	-	-	Nitrogen oxides (NOx). Carbon monoxide (CO). Carbon dioxide (CO2). Thermal decomposition can lead to the release of irritating gases and vapours.	https://www.fishersci.com/store/msds?partNumber=AC156445000&productDescription=3%2C5-DINITROSALICYLIC+ACI+500GR&vendorId=VN00032119&countryCode=US&language=en
Sodium potassium tartrate	-	-	-	Irritation, nausea, headache	-	Forms carbon oxides and sodium oxides upon degradation.	https://www.fishersci.com/store/msds?partNumber=S386500&productDescription=SOD+POT+TARTRA+USP+500GM&vendorId=VN00033897&countryCode=US&language=en
Sodium hydroxide	Skin corrosion/irritation	Serious damage/irritation	Ingestion causes damage to delicate tissues and can cause perforation	Acute toxicity to the respiratory system	Not flammable	Do not empty drains as high pH will harm aquatic life	https://www.fishersci.com/store/msds?partNumber=BP359500&productDescription=SODIUM+HYDROXIDE+500G&vendorId=VN00033897&countryCode=US&language=en
Nitric acid	Skin corrosion/irritation	Serious damage/irritation	Ingestion causes severe swelling, severe damage to the delicate tissue and danger of perforation: The product is a corrosive material. The use of gastric lavage or emesis is contraindicated.	Acute toxicity to the respiratory system. Corrosive to respiratory organs.	May intensify fire	Oxidizes metals. Do not let it enter drains. Low pH can affect aquatic life.	https://www.fishersci.com/msds?productName=A467250%26productDescription=NITRIC
Millon's reagent	Acute dermal toxicity. Skin corrosion/irritation. Fatal in contact with skin.	Serious damage/irritation	Acute oral toxicity	Acute inhalation toxicity	-	Toxic to aquatic organisms. Do not allow to enter groundwater systems.	https://www.fishersci.com/store/msds?partNumber=AC458541000&productDescription=MILLON+S+REAGENT%2C+FOR+TH+100ML&vendorId=VN00032119&countryCode=US&language=en
Sodium nitrate	Skin irritation	Serious damage/irritation	Acute oral toxicity. Toxic if swallowed.		May intensify fire	Very toxic to aquatic organisms.	https://www.fishersci.com/store/msds?partNumber=S3383&productDescription=SODIUM+NITRITE+USP%2FFCC+3KG&vendorId=VN00033897&countryCode=US&language=en
Sulphanilic acid	Skin corrosion/irritation. Skin sensitization.	Serious damage/irritation	-	May cause trouble breathing and dizziness	-	Toxic to aquatic organisms. Do not allow to enter groundwater systems.	https://www.fishersci.com/store/msds?partNumber=A296500&productDescription=SULFANILIC+ACID+CERT+ACS+500G&vendorId=VN00033897&countryCode=US&language=en
sodium carbonate	-	Serious damage/irritation	-	-	Not flammable	Do not empty drains	https://www.fishersci.com/store/msds?partNumber=AC424280025&productDescription=SODIUM+CARBONATE%2C+ANHYDR+2.5KG&vendorId=VN00032119&countryCode=US&language=en

(Table 18.1) cont.....

Reagent Name	Effect on Human				Fire Hazard	Effect on Environment	For more Information
	Skin	Eye	Ingestion	Respiratory Tract			
Ehrlich's reagent	-	-	Acute oral toxicity. Toxic if swallowed.	-	-	Very toxic to aquatic organisms. May cause long-term effects.	https://www.fishersci.com/store/msds?partNumber=D7125&productDescription=P-DIMET+AMINBENALDEHYD+CR+25G&vendorId=VN00033897&countryCode=US&language=en
alpha naphthol solution	Acute dermal toxicity. Skin corrosion/irritation.	Serious damage/irritation	Acute oral toxicity. Toxic if swallowed.	Respiratory irritation	-	Thermal decomposition can lead to the release of irritating gases	https://www.fishersci.com/store/msds?partNumber=AC128191000&productDescription=1-NAPHTHOL%2C+99%2B%25+100GR&vendorId=VN00032119&countryCode=US&language=en
Hypobromite solution	Irritation permeates into the skin, and skin sensitization	Eye irritant	Hazardous if swallowed	Hazardous if inhaled	-	-	https://www.oxfordlabchem.com/msds/(S-08147)%20SODIUM%20HYPOBROMITE%20SOLUTION.pdf
Lead acetate solution	-	-	Hazardous if swallowed. It is carcinogenic. Repeated exposure causes damage to organs. It is suspected of damaging fertility.	-	Non-flammable	Very toxic to aquatic organisms. May cause long-term effects.	https://www.fishersci.com/store/msds?partNumber=L33500&countryCode=US&language=en
Sodium nitroprusside	-	-	Toxic if swallowed. Acute oral toxicity.	-	-	May cause long-term adverse effects on the environment.	https://www.fishersci.com/store/msds?partNumber=S350100&productDescription=SOD+NITROPRUSSIDE+CR+ACS+100G&vendorId=VN00033897&countryCode=US&language=en
Ninhydrin	Skin corrosion/irritation. Skin sensitization.	Serious damage/irritation	Toxic if swallowed. Acute oral toxicity.	May cause respiratory irritation	-	Do not empty drains.	https://www.fishersci.com/store/msds?partNumber=AC415721000&countryCode=US&language=en
Acetone	Repeated exposure may cause skin dryness or cracking	Serious damage/irritation	Toxic if swallowed. Repeated exposure can cause damage to organs	May cause dizziness	Highly flammable	Should not be released into the environment.	https://www.fishersci.com/store/msds?partNumber=A18P4&productDescription=ACETONE+ACS+4L&vendorId=VN00033897&countryCode=US&language=en
Sodium dodecyl sulfate	Toxic in contact with skin	Serious damage/irritation	Harmful if swallowed	May cause respiratory irritation	Highly flammable	Very toxic to aquatic organisms. May cause long-term effects.	https://www.thermofisher.com/document-connect/document-connect.html?url=https://assets.thermofisher.com/TFS-Assets%2FLSG%2FSDS%2FRF0342_MTR-NARF_EN.pdf
Ammonium persulfate (APS)	Skin corrosion/irritation. May cause skin allergies.	Serious damage/irritation	Harmful if swallowed	May cause allergy or asthma symptoms if inhaled. It may cause respiratory irritation.	Highly flammable	Very toxic to aquatic organisms. May cause long-term effects. It is also harmful to the environment.	https://www.fishersci.com/store/msds?partNumber=AAJ7632209&productDescription=AMMONIUM+PERSULFATE+ACS+10G&vendorId=VN00024248&countryCode=US&language=en
TEMED	Causes severe skin burns	Serious damage/irritation	Harmful if swallowed	May cause respiratory irritation. May cause dizziness	Highly flammable	Should not be released into the environment.	https://www.fishersci.com/store/msds?partNumber=BP15020&productDescription=TEMED+20G&vendorId=VN00033897&countryCode=US&language=en
Acrylamide	Skin corrosion/irritation. May cause skin allergies. Carcinogenic agent.	Serious damage/irritation	Toxic if swallowed	Harmful if inhaled	-	Harmful to aquatic life. It contains components harmful to the environment.	https://www.fishersci.com/store/msds?partNumber=BP1368100&productDescription=ACRYLAMIDE%2FBISACR+37.5%3A1+100GR&vendorId=VN00033897&countryCode=US&language=en
Glycerol	-	-	-	-	-	Should not be released into the environment.	https://www.fishersci.com/store/msds?partNumber=AAJ64719AK&productDescription=GLYCEROL%2C%2C+99.5%2B%25+250ML&vendorId=VN00024248&countryCode=US&language=en
2-mercaptoethanol	Fatal in contact with skin. Skin irritation. May cause skin allergic reactions.	Serious damage/irritation	Damages fertility. Causes damage to organs through repeated exposure. Toxic if swallowed.	Toxic if inhaled	Highly flammable	Do not empty in drains. Should not be released into the environment.	https://www.fishersci.com/store/msds?partNumber=O3446I100&productDescription=2-MERCAPTOETHANOL+R+100ML+IND&vendorId=VN00033897&countryCode=US&language=en

(Table 18.1) cont.....

Reagent Name	Effect on Human				Fire Hazard	Effect on Environment	For more Information
	Skin	Eye	Ingestion	Respiratory Tract			
Methanol	Skin corrosion/irritation. May cause skin allergies.	Serious damage/irritation	Toxic if swallowed. Repeated exposure can cause damage to organs.	Respiratory irritation	Highly flammable	Should not be released into the environment.	https://www.fishersci.com/msdsproxy%3FproductName%3DA412SK4%26productDescription%3DMETHANOL%2BACS%2B4L%2BSAFE-COTE%26catNo%3DA412-SK4%2B%26vendorId%3DVN00033897%26storeId%3D10652
Ethanol	-	Serious damage/irritation	-	-	Highly flammable	Should not be released into the environment.	https://www.fishersci.com/store/msds?partNumber=BP8202500&productDescription=96%25+ETHANOL+SOLUTION&vendorId=VN00033897&countryCode=US&language=en
Dithiothreitol	Skin corrosion/irritation. May cause skin allergies.	Serious damage/irritation	Harmful if swallowed	Respiratory irritation	-	May form combustible dust concentrations in air.	https://www.fishersci.com/msdsproxy%3FproductName%3DBP1725%26productDescription%3DDTT%2B5G%26catNo%3DBP172-5%2B%26vendorId%3DVN00033897%26storeId%3D10652
Dichromate	Skin corrosion/irritation. May cause skin allergies.	Serious damage/irritation	Harmful if swallowed. Causes diarrhoea, vomiting and nausea.	Respiratory irritation. May be harmful if inhaled	Highly flammable	Very toxic to aquatic organisms. May cause long-term effects.	http://tools.thermofisher.com/content/sfs/msds/2016gwi/30d8a37deebdcebb98b32d16140e2006_MTR-NATS_EN.pdf
Silver nitrate	Skin irritation	Eye irritation	Harmful if swallowed. Causes diarrhoea, vomiting and nausea. May irritate mucous membranes.	Respiratory irritation. May be harmful if inhaled	-	Very toxic to aquatic organisms.	http://tools.thermofisher.com/content/sfs/msds/2016gwi/93621_MTR-NATS_EN.pdf
Paraformaldehyde	Skin corrosion/irritation. May cause skin allergies.	Serious damage/irritation	Harmful if swallowed. Acute oral toxicity.	Respiratory irritation. May be harmful if inhaled. Suspected of causing cancer	Highly flammable	-	https://www.fishersci.com/store/msds?partNumber=O4042500&productDescription=PARAFORMALDEHYDE+R+500G&vendorId=VN00033897&countryCode=US&language=en
4-Chloro-1-naphthol	Skin corrosion/irritation. May cause skin allergies.	Serious damage/irritation	-	Respiratory irritation.	-	-	https://www.fishersci.com/store/msds?partNumber=AC109550050&productDescription=4-CHLORO-1-NAPHTHOL%2C+97%25+5GR&vendorId=VN00032119&countryCode=US&language=en
3-Amino-9-ethylcarbazole	-	-	Toxic if swallowed. May cause cancer		-	Should not be released into the environment	https://www.fishersci.com/store/msds?partNumber=AC147870250&productDescription=3-AMINO-9-ETHYLCARBAZOLE+25G&vendorId=VN00032119&countryCode=US&language=en
Phenazine methosulfate	Skin irritation	Serious damage/irritation	-	Respiratory irritation	-	Should not be released into the environment	https://www.fishersci.com/store/msds?partNumber=AC130160010&productDescription=PHENAZINE+METHOSULFATE%2C+1GR&vendorId=VN00032119&countryCode=US&language=en
Bicinchoninic acid	Skin irritation	Serious damage/irritation	-	Respiratory irritation	-	Should not be released into the environment	https://www.fishersci.com/store/msds?partNumber=AA4424704&productDescription=BICINCHONINIC+ACD+DISOD+SLT+2G&vendorId=VN00024248&countryCode=US&language=en
Folin-ciocalteau reagent	Skin irritation	Serious damage/irritation	-	Respiratory irritation	-	Should not be released into the environment	https://www.fishersci.fi/store/msds?partNumber=10191520&productDescription=500ML+Folin+%2526+Ciocalteu%2527s+phenol+reagent%252C+pure&countryCode=FI&language=en
Biuret reagent	Skin irritation. Causes severe skin burns	Serious damage/irritation	-	-	-	Should not be released into the environment	https://www.fishersci.com/content/dam/fishersci/en_US/documents/programs/education/regulatory-documents/sds/chemicals/chemicals-b/S25200.pdf
TritonX-100		Serious damage/irritation	Harmful if swallowed. Acute oral toxicity.	-	-	Toxic to aquatic life with long-lasting effects.	https://www.fishersci.com/msdsproxy%3FproductName%3DBP151100%26productDescription%3DTRITON%2BX-100%2B100ML%26catNo%3DBP151-100%2B%26vendorId%3DVN00033897%26storeId%3D10652
Copper sulfate	Skin irritation.	-	Harmful if swallowed. Acute oral toxicity.	-	-	Toxic to aquatic life with long-lasting effects.	https://www.fishersci.com/msds?productName=AC422870050

(Table 18.1) cont.....

Reagent Name	Effect on Human				Fire Hazard	Effect on Environment	For more Information
	Skin	Eye	Ingestion	Respiratory Tract			
Citric acid	-	Serious damage/irritation	-	Respiratory irritation	-	May form combustible dust concentrations in the air.	https://www.fisherscі.com/store/msds?partNumber=AC423565000&productDescription=CITRIC+ACID%2C+ANHYDROUS%2C+500GR&vendorId=VN00033901&countryCode=US&language=en
Ethidium Bromide	-	-	Harmful if swallowed. Suspected of causing genetic defects	Fatal if inhaled	-	Should not be released into the environment.	https://www.fishersci.com/store/msds?partNumber=BP1025&productDescription=ETHIDIUM+BROMIDE+5G&vendorId=VN00033897&countryCode=US&language=en
Formamide	Suspected of causing cancer.	-	May damage fertility. May damage the unborn child	May cause damage to organs through prolonged or repeated exposure	-	Should not be released into the environment.	https://www.fishersci.com/store/msds?partNumber=AAA1107636&productDescription=FORMAMIDE+99%25+500G&vendorId=VN00024248&countryCode=US&language=en
EDTA	-	-	May cause damage to organs through prolonged or repeated exposure	Harmful if inhaled	-	Should not be released into the environment.	https://www.fishersci.com/msdsproxy%3FproductName%3DBP1201%26productDescription%3DEDTA%2BDISODIUM%2BSALT%2B1KG%26catNo%3DBP120-1%2B%26vendorId%3DVN00033897%26storeId%3D10652
Xylene Cyanol	Skin irritation	Serious damage/irritation	-	Respiratory irritation	-	Should not be released into the environment.	https://www.fishersci.com/store/msds?partNumber=BP56510&productDescription=XYLENE+CYANOL+FF+10G&vendorId=VN00033897&countryCode=US&language=en
Diphenylamine	Toxic in contact with skin	-	Toxic if ingested. May cause damage to organs through prolonged or repeated exposure.	Toxic if inhaled	-	Very toxic to aquatic organisms. May cause long-term effects.	https://www.fishersci.com/store/msds?partNumber=AC423651000&productDescription=DIPHENYLAMINE+100G&vendorId=VN00032119&countryCode=US&language=en
Orcinol reagent	Skin irritation	Serious damage/irritation	-	Respiratory irritation	-	Do not empty in drains.	https://www.fishersci.com/store/msds?partNumber=AC416581000&countryCode=US&language=en
Sodium acetate	Toxic. May cause severe burns	Serious damage/irritation	Toxic if swallowed. Suspected of causing genetic defects. Causes damage to organs through prolonged or repeated exposure.	Toxic. Respiratory irritation. May cause drowsiness/dizziness.	-	Should not be released into the environment	https://www.fishersci.com/store/msds?partNumber=AC221750015&productDescription=PHENOL%2C+99.5%25%2C+EXTRA+PUR+1.5KG&vendorId=VN00032119&countryCode=US&language=en
Chloroform	Causes skin irritation	Causes serious eye irritation	Harmful if swallowed. Suspected of damaging the unborn child	Acute Toxic Toxic if inhaled. Suspected of causing cancer. May cause drowsiness or dizziness	-	Should not be released into the environment.	https://www.fishersci.com/store/msds?partNumber=AC610281000&productDescription=CHLOROFORM+ANHYD&vendorId=VN00033901&countryCode=US&language=en
guanidine hydrochloride	Causes skin irritation	Causes serious eye irritation	Harmful if swallowed	Harmful if inhaled	-	May form combustible dust concentrations in the air	https://www.fishersci.com/store/msds?partNumber=AAJ75823A1&productDescription=GUANIDINE+HYDROCLRIDE+99.5+1KG&vendorId=VN00024248&countryCode=US&language=en
Sudan IV stain	Causes skin irritation	Causes serious eye irritation	-	May cause respiratory irritation	-	Should not be released into the environment	https://www.fishersci.com/store/msds?partNumber=S66725&productDescription=sudan-iv-certified-biological-stain-fisher-chemical&vendorId=VN00033897&keyword=true&countryCode=US&language=en
Potassium hydrogen sulfate	Causes severe skin burns	Serious damage/irritation	-	May cause respiratory irritation	-	Should not be released into the environment	https://www.fishersci.com/store/msds?partNumber=P193500&productDescription=POTASSIUM+BISULFATE+CERT+500G&vendorId=VN00033897&countryCode=US&language=en

(Table 18.1) cont.....

Reagent Name	Effect on Human				Fire Hazard	Effect on Environment	For more Information
	Skin	Eye	Ingestion	Respiratory Tract			
Phenolphthalein	-	-	Suspected of causing genetic defects. Suspected of damaging fertility	Suspected of causing cancer	-	Should not be released into the environment	https://www.fishersci.com/store/msds?partNumber=P79500&productDescription=PHENOLPHTHALEIN+CERT+ACS+500G&vendorId=VN00033897&countryCode=US&language=en
Acetic anhydride	Causes skin irritation	Causes serious eye irritation	Harmful if swallowed	Fatal if inhaled	Flammable liquid and vapour	Should not be released into the environment	https://www.fishersci.com/store/msds?partNumber=AC222130100&productDescription=ACETIC+ANHYDRIDE%2C+98%2B%25+10LT&vendorId=VN00032119&countryCode=US&language=en
Potassium hydroxide	Causes severe skin burns	Causes serious eye irritation	Harmful if swallowed	May cause respiratory irritation	-	Very toxic to aquatic organisms. May cause long-term effects.	https://www.fishersci.com/msdsproxy%3FproductName%3DP2503%26productDescription%3DPOT%2BHYDROXIDE%2BCERT%2BACS%2B3KG%26catNo%3DP2503%26vendorId%3DVN00033897%26storeId%3D10652
Diethyl ether	-	-	Harmful if swallowed	May cause drowsiness or dizziness	Extremely flammable liquid and vapour	Should not be released into the environment.	http://www.thermofishersci.in/technical_msds/diethyl-ether.pdf
Iodine monochloride	Toxic in contact with skin. Causes severe skin burns	Causes severe eye damage	Fatal if swallowed	May cause respiratory irritation	-	Should not be released into the environment.	https://www.fishersci.com/store/msds?partNumber=AA4434814&productDescription=IODINE+MONOCHLORIDE+ACS+25G&vendorId=VN00024248&countryCode=US&language=en
Potassium iodide	-	-	-	Causes damage to organs through prolonged or repeated exposure	-	Should not be released into the environment.	https://www.fishersci.com/msdsproxy%3FproductName%3DAC418260050%26productDescription%3DPOTASSIUM%2BIODIDE%2BFOR%2BANALYSIS%26catNo%3DAC418260050%26vendorId%3DVN00032119%26storeId%3D10652
Copper nitrate	Causes severe skin burns	Causes severe eye damage	-	May cause respiratory irritation	May intensify fire; oxidizer.	-	https://www.fishersci.com/store/msds?partNumber=AC405855000&productDescription=COPPER%28II%29+NITRATE+2.5-+500GR&vendorId=VN00032119&countryCode=US&language=en
Trifluoroacetic acid	Causes severe skin burns	Causes severe eye damage	-	Harmful if inhaled	-	-	https://www.fishersci.com/store/msds?partNumber=A11650&productDescription=TRIFLUOROACETIC+ACID%2C+OPTIMA+L&vendorId=VN00033897&countryCode=US&language=en

In the case of:

Eye Damage

Protect the unexposed eye. Rinse with water for 15-20 mins. Remove contact lenses, if present and easy to do. Continue rinsing. Do not apply (chemical) neutralizing agents without medical advice. Take the victim to an ophthalmologist.

Respiratory Difficulties

Take the person to fresh air and make him comfortable for breathing.

Skin Damage

Wash immediately with lots of water (15 minutes)/shower. Do not apply (chemical) neutralizing agents without medical advice. Remove clothing while washing. Do not remove clothing if it sticks to the skin. Cover wounds with a sterile bandage. Consult a doctor/medical service. If burned surface > 10%: take the victim to a hospital.

Ingestion

Rinse mouth with water. Immediately after ingestion: give lots of water to drink. Do not induce vomiting. Do not apply (chemical) neutralizing agents without medical advice. Immediately consult a doctor/medical service. Call Poison Information Centre (www.big.be/antigif.htm). Ingestion of large quantities: immediately to a hospital. Take the container/vomit to the doctor/hospital.

CONCLUSION

Laboratory safety precautions are often neglected by young researchers. However, it is one of the most important sets of rules to be followed not just for personal safety but also for the safety of the whole working team, and for the environment as well. The SDS information related to chemical compounds is mentioned in all chemical bottles. It is a crucial step to check the hazards that can be caused while handling that material, be it known or unknown to the person using it. Therefore, one must take all necessary precautions while handling the same. Some chemical compounds can harm the environment, which must be processed before discarding. Nonetheless, SDS related to chemical compounds is also available online on various websites, but the user must ensure to check the latest information.

VIVA VOCE

1. What is SDS?

Ans. SDS or safety data sheet is a comprehensive document having information about a chemical ingredient or mixer hazards, physical and chemical properties, toxicity, handling, storing, and any other important details regarding the chemical.

2. What is OSHA?

Ans. OSHA or Occupational Safety and Health Administration is a regulatory body of the United States Department of Labour providing safe and healthy guidelines for the workplace.

3. How many sections are prescribed by OSHA for SDS? Which sections are not non-mandatory for making SDS?

Ans. There are a total of 16 selections prescribed by OSHA for SDS.

Sections 12-15 are non-mandatory parts of SDS.

REFERENCES

[1] Nations, U. *Globally harmonized system of classification and labelling of chemicals (GHS)*; United Nations: New York, NY, USA, **2011**. [http://dx.doi.org/10.18356/4255cc90-en]

[2] Codes, standards, and regulations. *Health Manag. Technol.,* **1992**, 153-186.

[3] Occupational safety and health administration. *Hazard Communication.,* **2023**. Available From:https://www.osha.gov/hazcom

SUBJECT INDEX

D

E

F

G

H

P

Q

R

S

T

U

W

X

Made in the USA
Las Vegas, NV
26 October 2026

c27b915d-e570-4dc1-bea8-59f62a53af8aR01